MYSTERY SHIP!

A History of the Travel Air Type R Monoplanes

EDWARD H. PHILLIPS

FLYING BOOKS INTERNATIONAL
211 Alpha Lane, Fleming Field
So. St. Paul, MN 55075

MYSTERY SHIP

A History of the Travel Air Type R Monoplanes

By Edward H. Phillips

Published as part of the
"HISTORIC AIRCRAFT SERIES"

FLYING BOOKS INTERNATIONAL, Publishers & Wholesalers
211 Alpha Lane, Fleming Field
South St. Paul, MN 55075
651-453-1875
G.E. Herrick, Publisher

Wings of Stearman
by Peter M. Bowers

The Staggerwing Story
by Edward H. Phillips

Mr. Piper And His Cubs
by Devon Francis

Piper: A Legend Aloft
by Edward H. Phillips

Beechcraft, Pursuit of Perfection
by Edward H. Phillips

Cessna, A Master's Expression
by Edward H. Phillips

Travel Air, Wings Over the Prairie
by Edward H. Phillips

Wings of Cessna, Model 120 to the Citation X
by Edward H. Phillips

Speed, the Biography of Charles Holman
by Noel E. Allard

The 91 Before Lindbergh
by Peter Allen

DH-88: The Story of DeHavilland's Racing Comets
by David Ogilvy

The New Ryan
by Ev Cassagneres

Of Monocoupes and Men
by John Underwood

The Stinsons
by John Underwood

T-Hangar Tales
by Joseph P. Juptner

The "Historic Aircraft Series" is published to document the careers and service of the men, women, and machines that gave America Wings.

Mystery Ship

A History of the Travel Air Type R Monoplanes

Flying Books International,
Publishers & Wholesalers.
211 Alpha Lane, Fleming Field
South St. Paul, MN 55075 U.S.A.

Library of Congress Cataloging in Publication Data
Phillips, Edward H.
Mystery Ship! : a history of the Travel Air type R monoplanes / Edward H. Phillips.
p. cm. -- (Historic aircraft series)
Includes bibliographical references and index.
ISBN 978091113929 52995
1. Airplanes--Design and construction--History.
2. Travel Air (Firm)--History.
I. Title. II. Series.
TL671.2.P48 1999
629.133'343--dc21

99-40372
CIP

Front cover illustration: *The five Travel Air Type R Mystery S aircraft: Top - R2001/NR614K; Middle - Italian R2005/11717; Texaco 13 R2004/NR1313; Shell Oil R2003/NR482N; Chevrolair R2002/R613K.*

Rear cover illustration: *Bobby Younkin taxies the Younkin Travel Air R replica at Oshkosh '95. Noel Allard photo.*

Printed and bound in the United States of America
Art Director, Noel Allard
Publisher, G.E. Herrick

TABLE OF CONTENTS

INTRODUCTION

When it burst upon the aviation scene at the National Air Races near Cleveland, Ohio in September 1929, the Travel Air Type R monoplane quickly established itself as a benchmark design in the history of American aeronautics. Sleek, agile, and fast, engineer Herbert Rawdon's gleaming red racer proclaimed the beginning of the end of the biplane era in the United States.

The product of a highly clandestine project at the Travel Air factory, the Type R was the secret weapon that Walter H. Beech intended to unleash at Cleveland against the best machines his commercial competitors and military aviation could offer. With veteran pilot and Beech's long-time friend Douglas Davis at the stick, the little monoplane easily brushed aside its adversaries and flew to victory in the free-for-all contest at an average speed of more than 194 mph.

In the wake of that significant achievement, the Type R became an overnight sensation. To defeated foes, and particularly to the U.S. Army Air Corps and Navy pilots who flew against it that day, the racer represented the wave of the future. The handsome monoplane soon became the focus of many articles in trade magazines of the time such as *"Aviation"* and *"Aero Digest,"* and Travel Air's parent company Curtiss-Wright Corporation was quick to capitalize on the racer's accomplishment.

In addition to the famed "Mystery Ship" that won the coveted Thompson Cup in 1929, four other Type R monoplanes were built by the Travel Air company. Of these the second was reengined and sold to flamboyant aviatrix Florence "Pancho" Barnes. The third airplane was operated by the Shell Oil Company and flown by James H. "Jimmy" Doolittle and James Haizlip, and the fourth machine was procured by The Texas Company for the express use of its superintendent of aviation, Frank Monroe Hawks. The fifth and final Type R was built to order for Italy.

More than any other airplane of its time, the Type R demonstrated that raw horsepower alone was not the most important ingredient in a successful air racing machine. It was the careful, deliberate blending of a lightweight airframe with a powerful engine that ultimately paved the way to the checkered flag.

Although there was nothing technically revolutionary about the Type R, it did represent a major evolutionary step in design philosophy. The semi-cantilever wing incorporated a thin airfoil and the slender fuselage was carefully streamlined to accommodate a bulky radial engine while minimizing drag. The combination provided superior performance compared with the most advanced biplanes of the era.

Conceived by Herbert Rawdon and brought to life with the help of talented co-engineer Walter Burnham, the legendary Type R turned the aeronautical world upside down. The little monoplane from Wichita forced commercial and military analysts and aircraft designers to break with tradition and begin to seriously embrace the monoplane and its inherent advantages.

More importantly, the airplane blazed a trail for others to follow. In the hands of skilled airmen such as Doolittle, Hawks, and Haizlip, the speedy Type R kept the name of Travel Air at the forefront of aviation for nearly three years. It brought lasting fame to an airplane company in Kansas and reinforced the growing reputation of its ebullient leader, Walter Beech.

Although these Travel Air monoplanes left a distinct and permanent mark on aviation history, eventually they were eclipsed by more sophisticated designs and gradually faded from the aviation scene. But not from the memories of those who experienced the speed and majesty that made the Type R unforgettable. *"Mystery Ship"* is the story of these famous airplanes and how they conquered the sky.

ACKNOWLEDGEMENTS

The author deeply appreciates the help rendered by many people during the research and preparation phases that led to the publication of this book. Without their assistance the story of the Travel Air Type R monoplanes could not have been written. My sincere thanks to all of them:

Kenneth D. Wilson, who modestly describes himself as "Largely a little-known aviation historian;" aero historians Peter M. Bowers, John W. Underwood; air racing experts Don Berliner and John Sunyak; Italian aviation historians Gianni Gambarini and Roberto Gentilli; Dan Hagedorn and the research staff of the Smithsonian Institution's National Air & Space Museum Library; Federal Aviation Administration; Raytheon Aircraft Company; Barbara Hunter Schultz; Bruce Bissonette; John Parish of the Staggerwing Museum Foundation; Faye Cox, Donald Baldwin and Arthur J. LeBlanc of The Texas Company; Sam Morton of the Shell Oil Company's "Shell And The American Landscape" Museum; the Kansas Aviation Museum; Rolf Norstog; and Mary Nelson of the Special Collections Department, Ablah Library, Wichita State University.

Birth of a Legend

CHAPTER ONE

Speed! Few men coveted it more than Walter H. Beech, president of the Travel Air Company in Wichita, Kansas. From his days as a fledgling pilot in the United States Army Signal Corps in 1919, Beech had developed an insatiable thirst for flying fast. A savvy entrepreneur as well as a talented aviator, by 1928 Walter had helped transform Travel Air from a small, obscure company to one of the world's premier airplane manufacturers. More importantly, he clearly understood the crucial role that performance played in selling airplanes. As a well-known aeronautical engineer once said, "People talk safety, but they buy speed." [1]

Air racing had always been one of Beech's favorite endeavors. He had initially earned his reputation as a pilot in the early 1920s when he flew Laird *"Swallow"* biplanes to victory, and in 1925 he had been victorious in the first National Air Tour for the Edsel B. Ford Reliability Trophy flying a Curtiss-powered Travel Air biplane. A year later, he and navigator Brice Goldsborough had won the 1926 Reliability Tour for the Edsel B. Ford Trophy in another Travel Air. As a seasoned competitor himself, Walter Beech recognized the importance of air racing, not only for its publicity value but also for its contribution to advancing airframe and engine technologies.

In the wake of the National Air Races in 1928 Beech was becoming increasingly unhappy that the competition was dominated, as it had been since the mid-1920s, by the powerful biplanes of the United States Army Air Corps and the United States Navy. With their streamlined aircraft powered by modified V-12 Curtiss *"Conqueror"* engines, the Army and Navy often drew the lion's share of public attention on each day of the races. Although commercially-designed airplanes normally did not compete directly against these pursuit machines, they would have stood little chance of winning.

Walter bristled at the thought of civilian flyers and their airplanes being upstaged each year by military aviators. To make matters worse, Travel Air had nothing to offer. Being second-best was anathema to Walter and ran counter to his tough, competitive spirit. One day after the 1928 races, a frustrated Beech lamented to his engineers

In 1928 the fastest airplane in Travel Air's stable was a Type D4000 fitted with speed wings and a 225-hp. Wright J-5 radial engine. Maximum speed was about 150 mph. Paul E. Braniff, later of Braniff Airlines fame, bought the aircraft illustrated in 1928. Assigned constructor number (c/n) 690 and registered C6239, it was among the earliest D4000s built. ***(Raytheon Aircraft Company)***

The rare Travel Air Type D2000, also known around the factory as the "Bug," featured special speed wings and a Curtiss OX-5 engine enclosed in a bulbous cowling. The single-seat ship was designed to compete in the 90-hp. class events but eventually was banned because of its superior performance. Maximum speed was about 125 mph. Note the narrow fuselage, wing fillets, fuel tank in the center section, and non-standard empennage shape. In 1929 X6473 was modified with a Wright radial engine to compete in the Women's Air Derby but crashed in Arizona, killing pilot Marvel Crosson. ***(Truman C. Weaver via Peter M. Bowers)***

Designed by Reginald J. Mitchell, the low-wing, externally-braced Supermarine S.4 racer influenced Herbert Rawdon's concept for a land-based, air racing monoplane. The S.4 illustrated (N-220) was flown by Sydney Webster to first place in the 1927 Schneider Trophy Race at a speed of 281.6 mph. The seaplane was powered by an 875-hp. Napier Lion VIIB and featured a wing span of 26 feet 8 inches and length of 24 feet 2 inches. Mitchell designed a series of racers for the Schneider contests, but he is chiefly remembered for creating the Supermarine "Spitfire" fighter of World War Two fame. ***(Vickers Ltd., E.B. Morgan via Don Berliner)***

that the fastest machine on Travel Air's flight line was a Model D4000 biplane equipped with "speed" wings and powered by a 225-hp. Wright J-5 radial engine. It could achieve a maximum speed of 150 mph. but it was hopelessly outclassed by the faster, more powerful Army and Navy fighters.

One of the engineers on the receiving end of Beech's displeasure was Herbert Rawdon. He not only shared Walter's viewpoint, but Rawdon also understood the problem and knew how to solve it. Instead of waiting until two months before the races and hastily modifying an existing airplane for the upcoming competition, as was Travel Air's custom, Rawdon realized that the company needed to initiate a dedicated, long-term project aimed specifically at creating an aircraft designed expressly for high-speed flight.

Not long after Beech's visit to the engineering department Rawdon was summoned to the front office where a spirited conversation was underway about air racing. Although Rawdon said little, he listened intently as Beech and some associates complained about the lack of a fast commercial airplane capable of doing battle with the military.

As Rawdon recalled years later, on the way home that evening in September he became convinced that work on a special Travel Air would have to begin almost immediately if the airplane was to be ready for the 1929 races. As his idea took root, a number of key issues became apparent. First, Rawdon could expect no help from Travel Air management because there was no money available to develop an air racer. Second, the engineering work would have to be done without pay and at home in his spare time. Rawdon knew that the only compensation he would receive for the Herculean effort that lay before him was the satisfaction of a job well done. Third, secrecy was imperative. No one, including Walter Beech, was to know of the project's existence until the time was right.

None of these restrictions, however, dampened Rawdon's enthusiasm. But he decided to enlist an able, trustworthy assistant to help him complete the design and development of the airplane within the next 12 months. His choice of fellow Travel Air engineer Walter Burnham proved to be providential. Burnham was a quiet man who rarely complained, and his competence with a slide rule and prowess with wood and sheet metal was beyond reproach. Their personalities would complement the project, not inhibit it.

In addition, Rawdon and Burnham were hands-on engineers that understood how to translate complex technical drawings into skeletons of steel tube, wood and fabric on the shop floor. Designing, developing, building and flight-testing a thoroughbred racing airplane in less than a year was a daunting task for an entire company, let alone only two eager engineers. Together, they would create an airplane that became an aviation benchmark in both form and function.

Within days Rawdon had completed some rough sketches of the airplane and discussed them with Burnham. He envisioned a low-wing monoplane with a wire-braced wing and fixed landing gear. Inspiration for this configuration stemmed from Herb's admiration of the sleek, high-powered floatplanes that competed in the Schneider Trophy Races of the 1920s and early 1930s.

Of these, Rawdon was especially influenced by the British-built Supermarine machines that included the graceful S.5 powered by a 875-hp. Napier Lion VIIB engine. Designed by Reginald J. Mitchell, whose Schneider racers formed the foundation for development of the legendary Supermarine *"Spitfire"* fighter of World War Two fame, an S.5 won the 1927 Schneider contest held in Venice, Italy at a speed of more than 281 mph. Another S.5 placed second at 273 mph.

The S.5 racers were elegant in appearance. Their narrow, slender fuselage was mated to a thin wing braced by wires and mounted on a massive superstructure that supported the large floats. Aerodynamic streamlining was evident throughout the airframe, with careful attention paid to every joint where two or more surfaces met. The engine was nestled in a tight-fitting cowl that mated smoothly with the fuselage, further minimizing drag. Mitchell's S.6-series racers built in 1929-1931 were recognized as a paragon of floatplane design and won the Schneider Trophy outright for Great Britain in 1931. In addition, that year an S.6 became the first airplane to exceed 400 mph. Rawdon also admired the sleek Italian racing floatplanes built by Macchi and Macchi-Castoldi for the Schneider competitions. Highly streamlined and fast, these airplanes were the equal of their British counterparts in the late 1920s and early 1930s. [2]

For Herb, however, it was the use of a V-type powerplant in the Supermarine and Macchi machines that played a key role in his eventual decision to power the new Travel Air with an in-line engine. His decision was a logical one, especially since it was well known that the in-line design presented less frontal drag than a static radial engine of equal horsepower. By keeping the racer's weight to a minimum consistent with requirements for structural integrity, and by reducing drag through careful design of the airframe, Rawdon calculated that an inverted, in-line engine of 250-300 hp. would be adequate to achieve a speed of at least 165 mph.

Because such a powerplant was not readily available in the United States during late 1928, Rawdon used the weight of a Wright J-4 radial engine (475 pounds) for his initial design calculations and speculated that a suitable six- or eight-cylinder in-line engine would become available in 1929. Both Rawdon and Burnham were familiar with the aging J-4, which had powered early Travel Air biplanes beginning in 1926 but had become obsolete with the advent of the more powerful, nine-cylinder Wright J-5-series engines.

Since the airplane could not carry an official designation of the Travel Air Company during the design process, Rawdon dubbed the airplane the "Type R" in honor of his own last name. With the basic configuration established Herb and Walter began detailed design work in mid-September. An inboard profile of the airplane was made and a weight and balance estimate completed. The latter indicated the airplane would have a maximum gross weight of about 1,750 lbs. [3]

To ensure that the overall structure would withstand the severe stresses associated with air racing, the air-

The first Travel Air to be dubbed "Mystery Ship" by the press was the "Special" designed by Lloyd Stearman and Mac Short. It was built in 1925 and was powered by a Curtiss C6A in-line engine rated at 160 hp. at 1,750 RPM. The coolant radiator, mounted forward of the landing gear struts, could be retracted slightly to reduce drag. The biplane's maximum speed was about 125 mph. ***(Raytheon Aircraft Company)***

frame was designed to a load factor of 9.0 in the high angle of attack flight condition (nine times the airplane's maximum weight). A load factor of 7.0 was chosen for the landing condition chiefly because the wire bracing would be attached directly to the landing gear. The entire wing and landing gear structures were co-dependent on the wire bracing for their structural rigidity, therefore both assemblies had to be designed to a relatively high load factor. At low angle of attack the load factor was 6.0, and for the nose dive and inverted flight conditions the load factor was 3.6. A safety margin of 33.3% was incorporated into the airframe structure to provide an ultimate load factor of 12.0—equal to requirements for Army and Navy fighters of the era.

Early in the project Rawdon decided that the stress analysis surveys should comply with the latest Department of Commerce airworthiness regulations for issuance of an Approved Type Certificate, especially if the Travel Air Company chose to place the Type R into production. All stress analysis computations were based on the *"Proposed Handbook for Airplane Designers"* published by the Department of Commerce early in 1928. It provided engineers with acceptable methods of distributing loads and analyzing structures of commercial aircraft designs that would be submitted to the Department for issuance of an Approved Type Certificate. In addition, the handbook was intended to supplement the Air Commerce Regulations that became effective on December 31, 1926. [4]

Wing Design and Construction

By late in 1928 Rawdon and Burnham had begun the laborious task of preparing detailed engineering drawings and computing stress analysis for the Type R racer. Although general design of the fuselage, empennage (tail section), and fixed landing gear had been determined by that time, the choice of an airfoil for the wing proved more daunting. At least six airfoils were considered, including the N-15, U.S.A.-51, C-62, the Eiffel-375, as well as the NACA M-3 and NACA M-6, and the RAF 34.

Photographed inside the experimental department at the Travel Air factory, R614K was undergoing final assembly when it was photographed in July 1929. In the background the fuselage for R613K was being welded. Security was strictly enforced in and around the shop environment and access was limited. **(Raytheon Aircraft Company)**

THE QUESTION OF WINGS

The dearth of detailed documentation on each of the five Type R monoplanes has created uncertainties about the construction and disposition of wings that were reportedly built for the airplanes. Although the Wright-powered ship was fitted with its original, standard set of wings having a span of 29 feet, two inches for the 1929 National Air Races at Cleveland, Ohio former Travel Air employee Theodore "Ted" Cochran told the author in 1981 that he helped build one set of "small wings" for the Wright-powered monoplane (R614K) before the races. He stated in another interview that the wings were fitted to the airplane and test-flown at the factory, but that they were removed and the standard wings reinstalled.

These short-span wings apparently were built, but there is no known evidence in Travel Air company records, in Herb Rawdon or Walter Burnham's engineering files, or from other reliable sources that reveals their exact dimensions or eventual disposition. According to Clarence Clark, Travel Air's chief pilot, the wings were shorter than the standard set for the Wright-powered Type R. In 1981 Clarence Clark said that he flew R614K with the short-span wings installed, but that the increase in speed was minimal and the panels were removed in favor of the standard wings.

But Rolf Norstog, who had researched the "Mystery Ships" during the 1950s and 1960s and interviewed both Rawdon and Burnham, claimed that the standard wings were removed from R614K before the Cleveland races and were placed in storage at the Travel Air factory. He contended that the short-span wings were installed and remained on the airplane throughout its career, but there is no evidence to support that position. To the contrary, in a letter written by historian William Fleming to Walter Burnham on October 3, 1962 Fleming asked Burnham about the existence of two sets of wings for R614K. In response Walter Burnham scribbled a note stating, "only one set of small wings, Doolittle had them, R614K didn't." Although Burnham's comment is open to interpretation, he apparently was implying that the short wings built for the Wright-powered Type R were eventually acquired by Doolittle and were not installed on the Wright-powered airplane.

In another interview Ted Cochran claimed that the racing wings for R614K eventually were acquired by Frank Hawks for use on "Texaco 13," the fourth Type R built. In the 1970s these wings, still carrying the Texaco logo and NR1313's paint scheme, were placed on display by the Staggerwing Museum Foundation in Tullahoma, Tennessee. The museum obtained the wings from antique airplane enthusiast David Jameson, who also claimed they were from R614K. Although this is possible, there is no known documentation to substantiate the Museum's or Jameson's position. [5]

Research by the author, however, indicates that James Doolittle may have obtained the wings from R614K in March, 1930 for occasional use on the third Type R, the Shell Mystery Ship, which was built before the Texaco 13. In addition, Doolittle wrote to the author in 1984 and stated that he had installed an "extra set" of wings on the Shell Mystery Ship when he had the wrecked airplane rebuilt to his specifications in 1931. Doolittle wrote that the wings "were from an old ship"—a possible reference to R614K (refer to Chapter Five).

Based on the available evidence, it appears that Doolittle, not Hawks, may have acquired the racing wings from R614K and that the airplane flew most of its career with a standard set of wing panels. Records of The Texas Company indicate that a separate set were later built for use on Texaco 13. [6]

In addition, a short-span set of wings for the Chevolair-powered Type R may have been built for Florence Lowe "Pancho" Barnes, who bought the airplane in 1930 after it had been reengined with a Wright J-6-7 powerplant. According to Rolf Norstog, the racing wings were placed in storage at H.C. Lippiatt's Travel Air dealership at Grand Central Airport in Glendale, California after Barnes took delivery of the airplane.

All five of the Type R monoplanes were built using identical techniques. The Shell Mystery Ship (c/n R-2003, NR482N), photographed during construction early in 1930, illustrates how the 1/16-inch plywood covering was applied over the fuselage. Note mounting points on the stub wing for landing gear struts. ***(Raytheon Aircraft Company)***

Of these, only two were given serious consideration and studied extensively—the M-6 and the RAF 34. The former was developed by the National Advisory Committee for Aeronautics (NACA) and had been used by Travel Air on the Type 5000 cabin monoplane of 1927. Rawdon derived information about the M-6 airfoil from NACA Report Number 244.

The RAF 34 was a British airfoil design that Herb had learned about in NACA Report Number 286. Both airfoil sections were similar in many respects and were considered adequate for the task. The M-6, however, initially was chosen because of its "favorable center of pressure movement, good spar depth, and good value of maximum lift and low drag," according to Rawdon's engineering notes. But after a lengthy, careful study the RAF 34 section was eventually selected, chiefly because Herb concluded that it "would be preferable to the M-6 from considerations of more favorable characteristics at the stall." He also observed that "this wing appears to be very efficient and if aerodynamic data could be obtained from a density tunnel, its characteristics would undoubtedly be as good as those of the M-6." The RAF 34 also had an acceptable lift-to-drag ratio. In addition, the center of pressure (for stress analysis purposes) of both airfoils as well as their planforms were "identical," according to Rawdon. As a result, the stress analysis

investigation he performed for both wings centered on the strength of their spars.

The Travel Air Type "B" wings (Frise ailerons) designed for the Type R featured an RAF 34 airfoil having a constant thickness from the inboard rib outboard to the attach points for the flying and landing wires, where the wing gradually tapered to the elliptical-shaped tip. The spars were fabricated from two, box-type spruce beams that were glued together but were not routed out internally. Ribs also were spruce, built up from 5/16-inch square strips with 1/16-inch mahogany plywood gussets at each joint. They were spaced along the spars at 12 1/4-inch intervals in the area where the wing was outside of the propeller slipstream (which accounted for 101.3 square feet of total wing area). By contrast, only 21.7 square feet of wing area was inside the slipstream region.

Compression ribs featured a heavy spruce truss that was boxed with plywood webs for additional strength. The ribs were reinforced where the flying and landing wires attached to the internal wing fittings using 3/4-inch maple blocks tapered and feathered on all sides. To further strengthen the spars at the outboard wire attach fittings, birch was incorporated into the outer two laminations. Drag wires, which stiffened the internal wing structure, were Macwhyte Number 10-32 round, swaged tie rods. All wing fittings were made from chrome molybdenum steel stock, and were designed for loads 20% above the level that the spars and other associated structural members were capable of withstanding. Bolts used to attach the fittings were heat treated to give an ultimate tensile strength of 125,000 pounds per square inch.

To provide torsional stiffness the entire wing was covered with 1/16-inch thick, 3-ply Haskelite mahogany plywood that was glued, screwed, and nailed to the ribs and spars from the leading edge to the rear spar on both the upper and lower surfaces. According to Rawdon, "The purpose of this covering is to strengthen the leading edge against the very high intensity local loadings that are known to occur on high-speed airplanes, and to make the whole wing as stiff as possible in torsion, which is quite important in a racing airplane." The plywood also helped to prevent vibration at the wing tips. Weight of both wings, including ailerons and external wire bracing, was estimated to be 255 pounds.

As design of the wing progressed, some of the original specifications were changed. For example, total wing area had been estimated to be 141 square feet but was revised later to 125 square feet. Similarly, wing chord (with the M-6 section) was originally 63 inches, but the chord was reduced to 60 inches when the RAF 34 was selected. Wing span, which had been set at 30 feet 6 inches, was reduced to 29 feet 2 inches. Aspect ratio (ratio of span to chord) was altered slightly to 6.26 from 6.25. One additional change centered on the angle of incidence, which was reduced to 1 3/4 degrees for the RAF 34 airfoil from 2 1/4 degrees for the M-6 airfoil (incidence is the angle at which the wing is attached to the fuselage). Wing dihedral was established at four degrees.

Stub Wing Structure

The two wing panels bolted to a "stub" structure that also served as a mount for the landing gear assemblies. The structure weighed 50 pounds and extended outward from both sides of the fuselage about 25 inches. Two support struts on each side of the fuselage joined with the stub assemblies and with the landing gear struts. The wings were attached to the structure with bolts and relied entirely on the wire bracing for their rigidity. To cool the engine oil, a tubular radiator having a surface area of 5.33 square feet and a cross section of 1.15 inches was fabricated from .040-inch aluminum alloy. It was about 48 inches in length and was installed in the left stub wing.

Stabilizers and Control Surfaces

The vertical and horizontal stabilizers, ailerons, elevator and rudder were of welded steel tube construction with stamped steel ribs welded in position. Each surface was covered with Flightex Grade "A" fabric that was stitched at close intervals on each rib. The leading edge of the horizontal and vertical stabilizers were covered with two layers of fabric across 30% of the chord line for additional reinforcement. During static testing the horizontal stabilizer withstood a load of 50 pounds per square foot, and the vertical stabilizer supported a load of 37.5 pounds per square foot.

The Frise-type ailerons were attached to a false spar by three inset hinges. The surfaces deflected up 35 degrees and down 25 degrees, whereas elevator travel was 37.3 degrees up and 26.5 degrees down. Movement of the horizontal stabilizer to control longitudinal trim was limited to 1 1/4 degrees up to 1 1/4 degrees down. The pilot adjusted stabilizer position through a manual trim device to reduce aerodynamic loads on the control stick. The vertical stabilizer was adjustable, but only the ground. Although the elevator and rudder were controlled through 1/8-inch diameter, 7x19 stranded steel control cables, push-pull torque tubes were used to control the ailerons. None of the control surfaces were balanced, either statically or dynamically (no cases of flutter are known to have occurred in any of the five Type R airplanes built).

Fuselage

To provide a reasonable compromise between accommodating the pilot comfortably while minimizing frontal area, the fuselage structure was conventional in design and featured a welded, truss-type configuration fabricated from Society of Automotive Engineers (SAE) grade chrome molybdenum 4130 steel tubing. The longerons were 7/8-inch in diameter having a wall thickness of .035-inch, but tubing used in the forward fuselage was 1-inch in diameter with a wall thickness of .049-inch to carry loads induced into the airframe by the engine.

Certain sections of the structure were heat treated for increased strength. According to Travel Air Company documents, the fuselage was built by Ed Libby, Ervin "Dutch" Krueger, and Andrew Bland working under the supervision of Howard Baccus. After transcribing dimensions from the engineering drawings to the shop

The Type R's landing gear assemblies were simple but rugged. Two oil-type shock absorbers were mounted on each gear along with four springs. No brakes were installed on R613K or R614K, but brakes were fitted to airplanes built for The Texas Company, the Shell Oil Company, and Italy's Regia Aeronautica. ***(Raytheon Aircraft Company)***

floor, the longerons and their supporting trusses were laid out and tack-welded together using acetylene gas welding equipment. After transferring the fuselage framework to saw horses, the remaining diagonal brace tubes were installed and welded in place. A special jig tool was used to keep the entire assembly in alignment throughout the welding procedure, which was done carefully one section at a time to minimize distortion. Wire bracing was not used to stiffen the structure.

From the cockpit aft the fuselage was covered with 1/16-inch plywood with fabric tape applied between the plywood panels to ensure a smooth surface. Sheet metal was used to cover the fuselage between the cockpit and the firewall.

Landing Gear

Rawdon's decision to use a fixed arrangement instead of a retractable system was logical considering the factors of cost, weight, complexity, and the state of technology at the time. Although retractable systems were available, it must be remembered that the Type R had to be designed and built at a minimum of expense but without sacrificing safety or performance. Since the airplane's fundamental design dictated the use of external bracing to support the wing, the fixed landing gear played a crucial role as an anchor point for the three flying wires on each wing panel. As a result, the fixed gear arrangement was considered mandatory.

In addition, Rawdon and Burnham were well aware that a carefully designed and faired fixed gear exacted a small penalty in drag compared with a retractable installation. More importantly, the fixed gear's disadvantage in drag was more than offset by its lower weight, lack of complexity, and the elimination of cumbersome subsystems. In view of these factors, a retractable system was not given serious consideration. The majority of commercial and military airplanes built in 1928-1929 featured fixed landing gears, including the larger airline transports such as the Ford Trimotor and the Boeing Model 80.

Travel Air had no experience with retractable systems and did not employ them on any of the company's products. By the early 1930s, however, retractable gear installations were becoming standard equipment on an increasing number of commercial designs such as the Boeing 247, and later the Douglas DC-1. Retractable gear also was installed on smaller commercial aircraft such as the Beechcraft Model B17-series of 1934.

Weighing only 110 pounds, the Type R's "chassis," as Rawdon referred to it, was constructed from SAE 4130 streamlined steel tubing and was heat treated for strength and durability. It consisted of three major structural members that formed an "N" strut that was bolted to fittings on the stub wing assembly. Another set of smaller tubes were welded together to form a mount for the wheels, which were made by Bendix with a diameter of 24 inches and a rim width of four inches (no brakes were fitted to the first two airplanes built). The gear was braced laterally by Macwhyte streamline steel wires installed between the wheels and the stub wing structure. These wires countered aerodynamic loads imposed on

View of the aft fuselage of the Italian Type R (R-2005) reveals tail skid shock absorber installation (bottom) and chain-driven stabilizer trim system (upper right) that was controlled from the cockpit. Lifting bar (lower right) retracted into fuselage. Note internal wood stringers that stiffened the plywood fuselage covering. ***(Raytheon Aircraft Company)***

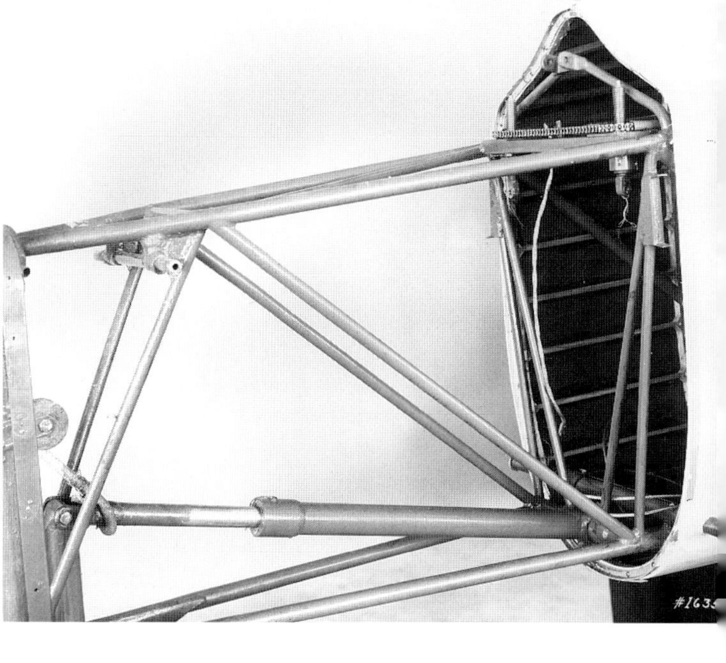

Detail view of cockpit floorboards in the Italian monoplane, R-2005. Control stick pivots are shown, as is the fuel strainer valve handle (lower right). ***(Raytheon Aircraft Company)***

the flying wires in flight.

To cope with stresses imposed by taxi, takeoff, and landing, a simple but effective set of two oleo-type shock absorbers were incorporated into each gear assembly. These used oil flowing through a metered orifice to cushion takeoff and landing loads. The units were capable of stroking up to 5.25 inches, and worked well on the rough airfields that the racer encountered during its career. The oleo units absorbed loads imposed during takeoff and landing, and a set off four springs on each gear handled taxi loads. The Type R's landing gear tread was 68 inches.

Overall, the unique landing gear design was elementary, efficient and rugged. Rawdon considered "that this type of shock is the most satisfactory of any to date for an airplane with such a high landing speed (70 mph.) and should aid very materially in preventing bouncing of the airplane on landing." The steel tail skid used a single oleo-type shock absorber with a stroke of 6 inches. The thrust line of the engine made an angle of 13 degrees with the horizontal when the skid was on the ground.

To reduce drag the wheels were covered by handmade speed fairings fabricated from .040-inch soft aluminum alloy. Carl Burnham, Walter Burnham's brother, was a master metal worker employed by Travel Air. He spent many hours hand forming and pounding the two halves of each fairing into shape to achieve the exact contours specified by Rawdon. The pieces were skillfully welded together along the centerline (by Carl Burnham) and were attached to the "N" strut assemblies. A small access panel on the outboard side of each fairing allowed mechanics to adjust tire inflation. The fairings were fixed to the gear structure and did not travel up and down with the wheels.

Wire Bracing

Installed between the wings, fuselage and landing gear were four sets of streamline steel tie rods that served as flying and landing wires. They were manufactured by the Macwhyte company. When tightened to proper tension the tie rods provided rigidity for the wing and landing gear assemblies and, therefore, were critical to maintaining the overall integrity of the wings, fuselage and landing gear assemblies that acted as a single structure.

Detailed photograph of the Shell Oil Company's Type R shows one of two fuel tanks installed in the forward fuselage between the support struts for the stub wing. Another tank was installed on the opposite side. The Texaco ship (R-2004) had a similar arrangement. The instrument panel was installed against flanged arch at left. Tubing for engine instruments is visible under left side of the arch. ***(Raytheon Aircraft Company)***

Above the wing two wires attached to the front and rear spar fittings at one end and the cabane strut fittings in the fuselage at the other end, whereas beneath the wing three wires were attached to a set of spar fittings but were anchored at the other end to the lower landing gear structure. Wood vibration dampers were installed on each set of wires.

Each of the 3/8-inch in diameter flying and landing wires installed on the wings were rated at 8,000 pounds tensile strength. At the forward spar, outer flying wire fitting the landing wire made an angle of 17 degrees with the wing spar and the flying wire made an average angle of 28 degrees. Similarly, at the rear spar outer flying wire fitting the rear landing wire made an angle of 17 degrees to the spar, and the rear flying wire made an angle of 28 degrees.

As for the landing gear, the two front tie rods were 3/8-inch in diameter each rated at 8,000 pounds tensile strength. The two rear wires, however, were 1/2-inch in diameter and each was rated at 15,500 pounds tensile strength. Because of their critical role in safety of flight, the wires were designed with margins of safety that varied from 30-60% above design load for the flying and landing wires, and 30-58% for the landing gear wires.

Two additional tie rods were installed on the empennage assembly to brace the horizontal and vertical stabilizers to each other, and another two braced the horizontal stabilizer to the lower fuselage longerons. These tie rods were 1/4-inch in diameter and each was rated at 3,400 pounds tensile strength.

Engine and Cowling

All of Rawdon and Burnham's initial stress analysis,

weight and balance, and performance computations for the Type R were predicated on acquiring an in-line engine of 250-300 hp. As 1928 gave way to the fateful year of 1929, the two men were continuing to work on detailed engineering drawings and a myriad of other chores associated with designing a special airplane. The painstaking work was not only hard on the men, but also on their families. Burnham's wife, Dorothea, told the author in an interview that Walter was away from home almost every evening working on the Type R project with Rawdon. One day, however, Walter came home before supper. As he later recalled, "One of the youngsters spotted me sitting in the easy chair and ran to his mother in the kitchen shouting, "Mommy, there's a stranger in the house!" 'But that's the way we worked back in those times, and that is how you can start from scratch and produce three airplanes in 10 weeks!"

Herb and Walter were making good progress, but the nagging problem of a suitable powerplant had to be resolved soon or the project would grind to a halt. As the warmth of spring spread across Kansas, in May 1929 an engine designer visited Travel Air and proposed that the company build a racing airplane to complement a 300-hp. six-cylinder, in-line engine that was under development.

Puffing on his ever-present pipe, Walter Beech listened intently and liked what he heard. Sensing an opportunity, Beech's enthusiasm grew by the minute. Fortunately, he had invited Rawdon to sit in on the discussion. After the meeting had concluded and the visitor had departed, Beech sought Herb's opinion of the proceedings. Without hesitation Rawdon revealed a set of three-view drawings depicting the Type R powered by an in-line engine.

The timing was perfect. Walter immediately grasped the little racer's potential if it could be mated to the proposed engine. As Rawdon later remembered, "Beech's enthusiasm took immediate effect, and he was all-out to build the project." But it was not to be. A few weeks later it became obvious to Beech that the engine designer had been overly optimistic and could not support his claims. The powerplant's design actually was in a state of infancy, not full-scale development. Both the would-be manufacturer and his "paper" engine quickly disappeared from the scene. [7]

The setback, however, did not affect Walter's determination to proceed. The 1929 racing season was fast approaching, and it was clear to Beech that Rawdon's design was exactly what Travel Air needed to reclaim its racing heritage. The thought of the little monoplane flying to victory in the prestigious National Air Races quickly elevated Walter's fierce, competitive spirit to a fever pitch. While Rawdon pondered where to find an engine, Beech made certain that design work on the Type R continued unabated and that it remained a closely-guarded secret.

There was more, however, for Beech to think about than air racing. At this point in Travel Air's five-year existence, it was about to become involved in the merger of two aviation giants—the Wright Aeronautical Corporation and the Curtiss Aeroplane & Motor Company. In 1928 the Wright company had acquired a

THE WRIGHT J-6-9 *"WHIRLWIND"*

In May 1927 the Wright J-5 static, air-cooled radial engine became the most famous powerplant in the world when it propelled Charles Lindbergh's Ryan monoplane dubbed the *"Spirit of St. Louis"* across the Atlantic Ocean from New York City to Paris, France. More importantly, however, the epic 33 1/2-hour flight demonstrated the inherent reliability of the air-cooled, static radial engine.

Although the J-5 Whirlwind proved to be a highly successful product, in 1927 the Wright company already was contemplating its replacement. By 1928 engine design was progressing at a rapid pace, and despite the J-5's performance it no longer made economic sense to build a nine-cylinder, 220-hp. radial when the same power could be achieved in a seven-cylinder configuration such as the new J-6-7 engine.

As a result, Wright engineers designed an entirely new, nine-cylinder powerplant designated as the J-6-9 (R-975) that developed 300 hp. at 2,000 RPM. It featured a 5:1 compression ratio, larger camshaft ring, and the pushrods and magnetos were mounted behind the cylinders, not in front as with the J-5-series engines. The design also incorporated automatic valve rocker arm lubrication, and many (but not all) of the engines had the exhaust manifold located at the front. Fuel consumption was about .55 pounds per horsepower per hour. The engine was 41 7/16-inches in length and had a diameter of 45 inches. It cost about $4,800 and weighed 520 pounds. Two Scintilla magnetos and a single-barrel Stromberg carburetor were standard equipment.

Introduced in 1929, the J-6-9 had a 5-inch bore and a 5 1/2-inch stroke. The cylinder barrels were steel with an

Wright J-6-9 "Whirlwind" engines powered four of the five Type R monoplanes. The photograph illustrates the powerplant with inter-cylinder baffles installed, but the original engine that powered R614K did not feature these baffles. ***(Smithsonian Institution print WA-1196)***

aluminum alloy head that was screwed and shrunk into place. Aluminum alloy pistons with full-floating, hollow pins were installed. Both the intake and exhaust valves were tulip-shaped, with the intake valves having solid stems and the exhaust valves hollow stems. A two-piece, single-throw crankshaft was used along with a one-piece master rod and "H" section articulated rods. The aluminum alloy crankcase was cast in four sections, and the camshaft follower housing with its tappet guides was cast integral with the main section of the crankcase.

During 1928, 1929 and 1930 the Travel Air Company bought a large number of these engines to power its Model 6000B (later Model 6B) cabin monoplane, as did the Cessna Aircraft Company to equip its DC-6A airplanes that were the equivalent of their Travel Air competitors. The 300 hp. engine also was installed on the Travel Air Model B9-4000 biplane as well as the rare, four-place Type 10B cabin monoplane. By 1931, however, the J-6-9 had been improved and was developing up to 450 hp. and could be modified to produce nearly 475 hp. Many other aircraft companies also used the engine and it eventually earned a reputation for dependability that exceeded its famous J-5 predecessor.

major interest in Travel Air, and the Wichita airplane builder had become another subsidiary along with the Loening and Keystone aircraft companies. The marriage of the Wright and Curtiss firms was tentatively scheduled to occur in August, and Walter Beech was destined to play an important role in that transaction.

Talk of planned mergers, however, did not help Rawdon find an engine for his airplane. Since there were no suitable in-line engines available he began considering the new, 300-hp. Wright J-6-9 (R-975) radial engine that would equip Travel Air's new Model 6000 cabin monoplanes. Pratt & Whitney also had a radial engine in the 300 hp. class and it also was under consideration. In view of the time constraints facing them, Rawdon discussed the situation with Beech and the two men decided that the J-6-9 was the best candidate engine.

The engine compartment of the Shell Mystery Ship illustrates a typical installation for the Type R airplanes. One of two magnetos is visible at the center below the shaft for the hand-cranked inertia starter. The Stromberg updraft carburetor, throttle control shafts, and Wright J-6-9 engine also are visible. Grease fittings were installed on valve rocker covers. The firewall was 0.040-inch aluminum alloy. ***(Raytheon Aircraft Company)***

As May evaporated into June, two key developments occurred that would profoundly affect the Type R project. Walter Beech had talked with Guy Vaughan, vice president of the Wright Aeronautical Corporation, about the possibility of the company providing a more powerful, high-compression version of a R-975 *Whirlwind* radial engine for a special application. Vaughan listened. Beech wanted at least a 400-hp. version, and Vaughan agreed to investigate the situation and report back to Walter. He soon contacted Beech and informed him that the company would cooperate to the fullest extent possible.

Walter quickly ordered Rawdon to travel to the Wright company's facilities in Paterson, New Jersey to discuss with factory engineers technical issues pertaining to the nine-cylinder powerplant. During his visit he learned that the standard J-6-9 engine would be modified to produce up to 425 hp. at 2,400 RPM., chiefly by increasing the supercharger ratio to 10.15:1 and making other technical changes to optimize performance. Compression ratio would be increased to 6.5:1 from 5:1. By comparison, a standard J-6-9 was rated at 300 hp. at 2,000 RPM.

When he returned to Wichita, Rawdon briefed Beech on the engine installation and a few days later Walter authorized construction of the racer. At last, Herb Rawdon had his engine. But it came in the form of a bulky, 580-pound air-cooled radial, not the compact, narrow in-line powerplant he had planned to use. He knew the J-6-9's large frontal area would present installation and drag issues that had to be resolved quickly—less than three months remained before the National Air Races would begin in Cleveland, Ohio.

One other significant development that directly affected the Type R project was the development in 1928 of the enclosed, pressure cowl by the National Advisory Committee for Aeronautics. The cowling program had begun at NACA's Langley Memorial Aeronautical Laboratory near Hampton, Virginia in June 1926. After more than two years of extensive experimentation that included work performed in Langley's new Propeller Research Tunnel, in November, 1928 the NACA published Technical Note 301 entitled *"Drag and Cooling with Various Forms of Cowling for a Whirlwind Engine."* Rawdon was familiar with this publication.

Confident that they had designed a cowl that both

cooled the engine cylinders adequately while significantly reducing drag, NACA engineers led by the talented Fred E.Weick borrowed a Curtiss AT-5A biplane from the United States Army Air Service and installed the new cowling around the Wright J-5 radial engine for flight testing. The results were impressive. Maximum speed increased to 137 mph. from 118 mph.—an increase of 16%, and engine cooling was improved.

The effectiveness of the NACA cowl was demonstrated to the public in February 1929 when pilot Frank M. Hawks flew a parasol-wing Lockheed *"Air Express"* from Los Angeles to New York nonstop in a record 18 hours, 13 minutes. The engine was fitted with an NACA cowl, which increased speed to 177 mph. from 157 mph. Gerald Vultee, at that time an engineer with the Lockheed Aircraft Company, sent a telegram to the Committee extolling the cowl's virtues and that the record would have been impossible without it. In 1929 the NACA won the prestigious Collier Trophy for its developmental work with the cowling. [8]

It is interesting to note that at the same time the NACA cowl was being developed in the United States, the Townend ring was invented by Hubert C. Townend at the British National Physical Laboratory. The circular ring, however, surrounded only the top portion of the cylinders, not the entire engine.

Although Beech was pleased that Wright Aeronautical was now involved in the racer project, he became increasingly concerned that the radial engine would create too much drag. Rawdon, however, assured him that by using the NACA's cowl design and modifying the outside diameter of the fuselage to mate smoothly with it, the drag penalties would be minor. Beech, who was not an engineer but had a keen appreciation for aerodynamics, accepted Herb's explanation.

Herb Rawdon swings the propeller to start the 420-hp. Wright J-6-9 radial engine installed on R614K. The photograph was made after the airplane emerged from the factory early in August 1929 but before the NACA-type cowling had been fitted. The letter "R" (Restricted license) had not been applied to the undersurface of the left wing. Note installation of exhaust stacks that mounted on the front of each cylinder. ***(Clarence Clark)***

Almost simultaneous with the decision to use the Wright J-6-9, Arthur Chevrolet, president of Chevolair Motors, Inc. of Indianapolis, Indiana and one of the Chevrolet Brothers of automotive fame, introduced an inverted, in-line six-cylinder aircraft engine rated at 170 hp. Air-cooled, the *Chevolair* D-6 was inadequate for the

Test pilot Clarence Clark keeps the Wright engine idling nervously as mechanics inspect the left elevator panel. The photograph was made soon after the airplane was rolled out of the factory for initial flight tests. The small windshield was later enlarged to provide the pilot with improved protection. Note wing root fairings and bright oil cooler tubing. ***(Clarence Clark)***

Type R but Walter Beech decided to build a second monoplane racer powered by the engine to compete in class events for smaller cubic inch-displacement engines. The airplane became the second Type R built for the Cleveland races. Beech also authorized a third racer to round out Travel Air's trio of competitors for the 1929 racing season. It was a diminutive biplane dubbed the Model B11D and was powered by a seven-cylinder, 240-hp. Wright J-6-7 radial engine. The airplane would be fitted with "speed wings" and a NACA cowling.

Modifications and Construction

With time working against them, in June and July 1929 Rawdon and Burnham labored feverishly to adapt the rotund Wright radial to the Type R's narrow airframe. The fuselage was enlarged and expanded forward of the cockpit to equal the diameter of the engine and the NACA cowling. A flurry of new stress analysis and weight and balance computations were made, along with revised engineering drawings and performance estimates. A new engine mount and ring also were designed for the radial powerplant.

In addition to these tasks the two engineers had to decide what parts to order for each airplane, determine when they had to arrive at the factory, and hope that their reasoning was correct. Procuring all of the necessary components to build three racers in 80 days seemed an insurmountable challenge. But as Rawdon later wrote, "This was not an easy problem, but with a little ingenuity and occasional substitution, the project made the grade."

Clarence Clark posed with a Travel Air Type BH biplane in 1925, not long after he joined the company as a pilot and mechanic. The 150-hp. Wright-Hispano Suiza engine features extended exhaust stacks that routed fumes away from the cockpit. Mechanical brakes were fitted to the landing gear. The Curtiss twisted-steel propeller was a new innovation that proved popular with aircraft manufacturers. The first airplane Clarence flew on a production test flight at Travel Air was c/n 10, a Model A biplane. Clark resigned from Travel Air in 1930 to fly for Phillips Petroleum Company in Bartlesville, Oklahoma. He died in 1988. ***(Raytheon Aircraft Company)***

Perhaps one of the most important decisions that had to be made centered on the choice of a propeller for the Wright-powered racer. Performance calculations originally made for the in-line engine were no longer valid, and new estimates indicated that the airplane would have a maximum speed of at least 205 mph. Based on his data, Rawdon contacted the Hamilton Standard propeller company in Milwaukee, Wisconsin and placed a special order on July 5, 1929. Specifications called for a maximum diameter of 8 feet 6 inches to allow for adequate tip clearance in the takeoff attitude, and that the propeller "must develop the maximum possible efficiency" at 200 mph., Rawdon wrote on the order form. He also was concerned about adverse effects of the large diameter cowling compared with the relatively small propeller, and he admonished Hamilton Standard to use "all means possible in design of the propeller to reduce fuselage/propeller interference."

Many years later Rawdon recalled a humorous incident that occurred between himself, Burnham and Walter Beech concerning the propeller for the Type R. About two weeks before the races were to begin, propellers for the Type R and the B11D biplane arrived at Travel Air. The shipping crates were opened and the propellers examined carefully before they were removed and placed on the floor. At noon that day Beech, who paid frequent visits to the experimental shop where the airplanes were being built, walked in for his customary check on progress.

Noticing the propellers, he commented that the 9-foot diameter propeller obviously was destined for the Type R, and the 8-foot propeller for the B11D. Walter Burnham, who overheard Beech's statement, responded

The Wright-powered Type R (R614K) flies above the Kansas Plains on an early test flight east of Travel Air Field in August 1929. Clarence Clark is at the controls. The racer had a maximum speed of 185 mph. without the cowling. After the cowl was installed speed at full throttle increased to 225 mph. Note grass streaming from the flying wires beneath the wings. Orthochromatic film made the red paint on the fuselage appear as dark as the black accent trim. ***(Raytheon Aircraft Company)***

that his assumptions were incorrect and explained that exactly the opposite was true. As Rawdon remembered it, "Somehow this did not seem logical to Beech, who became slightly agitated and immediately drew the conclusion that engineers were a confused group of individuals continually making bum mistakes in judgment." A brief but civil verbal exchange ensued but Beech remained unconvinced and departed in a huff, "muttering about the inconsistency of the average engineer." Of course, Rawdon and Burnham had made the correct choices.

With materials and component parts on order it was time to begin recruiting a work crew and begin construction. Only 10 weeks remained before the National Air Races, and there was no room for error in the coming rush to build three airplanes. Since they knew their co-workers well, Rawdon and Burnham selected the most talented and conscientious men for their special team, men who could work without supervision and do the job right the first time. A group of 25 workers was hastily assembled and briefed on the Type R project.

As work began early in July, Walter Beech had ensured that security was strictly enforced both inside and outside of the factory's experimental shop. Each worker pledged his silence. Thick canvas curtains were draped around the area, windows were covered to discourage the curious, and access to the shop floor was closely monitored. Beech knew he had a winner in the Type R, and he savored plans to surprise everyone at Cleveland with Rawdon's little speedster. Within days, however, the news had leaked out that Travel Air was "up to something big." Local newspaper reporters soon gathered at the factory on East Central Avenue, probing officials for information. None was forthcoming. As a last resort, reporters climbed ladders in an attempt to peak into the experimental shop in Building C, but the glass had been covered. The Wichita Eagle eventually reported that some type of "mystery ship" was under construction at the Travel Air facility, and the term soon became the unofficial moniker of the airplane.

The clandestine nature of the Type R project was

Profile view of R614K illustrates the airplane's sleek lines and compact design. The photograph was made before the Cleveland air races. The trailing edge of the windshield has been raised and the chord has been increased slightly. Panchromatic film made the dark red paint on the fuselage appear lighter than the black. ***(Raytheon Aircraft Company)***

noted in the August issue of *"Travel Air Currents,"* a factory publication for employees and customers: "Funny business is going on around the engineering and experimental departments. Mysterious packages and boxes are being delivered at odd hours. Groups of engineers and workers can be seen huddled here and there holding 'skull practice,' immediately dispersing upon the approach of an outsider. We have put the question fairly and squarely up to Herb Rawdon—asked him what was going on so we might pass the 'low down' on to you. And from him we got no answer except that perhaps in the next issue of *"Currents"* he would give us a peep. Have any idea what it is? Well, so do we, but we're not telling." The next issue, however, would not appear until after the Cleveland races had concluded.

According to Travel Air production records kept by factory manager William Snook, the three racing air-

Rear, quarter view of R614K illustrates black scallop accent trim on the vertical stabilizer. The three Type R ships built for the Shell Oil Company, The Texas Company, and the Italian government featured vertical stabilizers of increased area to improve longitudinal stability. ***(Kenneth D. Wilson)***

In this view of R614K the "R" has recently been applied to the upper surface of the right wing and the lower surface of the left wing. The letters may have been painted in a color other than black or a darker shade of red that created a sharp contrast when exposed to panchromatic film that was used to make the photograph. It also is possible that the letter was applied using a mint green color. The "R" on the rudder also contrasts the 614K registration on some photographs. Note the high gloss finish on the fuselage and wings. The airplane is parked outside of the Travel Air factory before departing for Cleveland, Ohio late in August 1929. ***(Kenneth D. Wilson)***

planes being fabricated that summer were not all assigned constructor numbers (c/n) in the normal Travel Air production sequence. The Wright-powered Type R (registered R614K) was designated as c/n R-2001, the Chevolair-powered ship (registered R613K) was c/n R-2002, and the B11D biplane (registered R612K) was assigned c/n 1267 in the factory's production sequence. Rawdon's personal designations, however, were different. He often referred to c/n R-2001 as R-101, and c/n R-2002 as R-102. These two airplanes were completed by workers in the experimental department during the week of August 13, 1929 and prepared for their initial test flights.

The original Wright J-6-9 installed on R614K was serial number 10688. As planned, it had been specially modified by Wright Aeronautical Corporation by removing the stock pistons and supercharger parts and installing high-compression pistons, rings, and high-speed supercharger gears that increased the blower ratio to 10.15:1.

Paint Design (R-2001, R614K)

Although the red base color and black and green trim colors applied to R614K are well known, detailed information about the original design of the black trim is lacking, especially as it pertains to the black scallops painted on the wings and empennage. The upper surfaces of each wing panel had large scallops applied to the stub wing section that decreased in size along the initial portion of the leading edge. Photographs of the airplane taken after construction do not indicate that scallops were applied to the underside of the wing leading edge. This practice was not unusual because paint added weight, and weight detracted from performance. In addition, it was more important that the upper surface of the wings be attractive than the bottom, which were less visible to observers. Based on photographic and documentary evidence, the author has concluded that in its original configuration R614K did not feature scallops along the entire span of each wing, either on the upper or lower surfaces, nor were they applied along the entire span of the horizontal stabilizers.

At least one photograph, however, taken in 1930 when the airplane was operated by the Curtiss-Wright Exhibition Company, reveals a pattern of scallops on the lower surface of the right wing. These probably were applied in May when the wings were recovered. Air racing historian John Sunyak, who had investigated appar-

Right, front quarter view of R614K reveals "R" applied under the left wing panel that contrasts black registration number. The cowling has been reinforced with additional fasteners to prevent aerodynamic forces in flight from shifting the unit forward. Slight damage to the left wing leading edge has been repaired and painted using silver dope, but finish color had not been applied. Of the five Type R ships built, only R614K had the pitot tube installed on the right wing. ***(Raytheon Aircraft Company)***

View of R614K reveals additional fasteners on cowling and race number "31" that has been applied to both sides of the fuselage. "R614K" painted on rudder is nearly invisible, possibly because of the type of film used to make the exposure or the angle of the rudder. ***(Raytheon Aircraft Company)***

ent discrepancies in R614K's scallop design, stated that in 1929 aviation artist Charles Hubbell and others were given access to the racer as a reference for their paintings. Some of these renderings were used as guides for detailing models of the popular airplane made by hobby vendors such as the Cleveland Model and Supply Company.

Sunyak photographed Hubbell's original paintings of R614K that were housed in the Crawford Museum in Cleveland, Ohio. One of these, dated sometime in 1932, showed two scallops on the undersurface of the right wing that were the same size and applied across the entire wing leading edge. Another painting by Hubbell shows two scallops evenly spaced on the upper surface of the right wing. But a drawing of the Type R copyrighted in 1935 by the Cleveland model company portrays the monoplane with five scallops on the upper and lower surface of each wing.

◀ *A detailed view of R614K's empennage shows fairing that deflected with the rudder. Note the use of strap hinges on elevators and rudder.* ***(Smithsonian Institution negative number 80-14857)***

Rear view of R614K vividly illustrates how the forward fuselage was widened to accommodate the bulky Wright J-6-9 radial engine. Black scallops are visible on the horizontal stabilizers, and the letter "R" on the right wing appears to be a slightly different shade of color than "614K." ***(Raytheon Aircraft Company via Mike Madewell)***

Front view of the airplane reveals the absence of inter-cylinder baffles on the Wright J-6-9 engine. Note that the left wing leading edge had been repaired but not repainted. **(Mike Madewell)**

The Type R's structural integrity in the air depended on proper tension of the steel Macwhyte flying and landing wires installed between the fuselage, wings, and landing gear. The firewall cowl aft of the NACA-type cowling is visible as well as the bonnet fairing that was fitted over the front of the engine for cylinder cooling and drag reduction. **(Smithsonian Institution negative number 80-14856)**

actual airplane at the 1929 race, why does one show two scallops and the other five?" He and other historians have speculated that the difference in the wing scallops was attributable to the set of short-span racing wings, but there is no known evidence to indicate that R614K was fitted with these wings during or after the 1929 National Air Races. [9]

Initial Test Flights Of R614K

Throughout the hot, humid Kansas summer the team of skilled craftsmen had worked long, hard hours to build Herb Rawdon's racers. Construction proceeded rapidly, and by mid-August the Wright-powered Type R was completed and ready for test flights. The NACA cowling, however, would not be available for installation until later that month. Despite that setback, Rawdon was anxious to fly the airplane to perform preliminary tests and collect data. Only two weeks remained before the airplane would have to depart for Cleveland, Ohio.

The choice of who would make the first flight was academic—Clarence E. Clark. He had been a test pilot for Travel Air since 1925 and by 1929 had worked his way up through the ranks to become chief pilot. Born in Iola, Kansas Clarence had learned to fly in 1923 from the back seat of a Curtiss JN-4 biplane powered by the ubiquitous Curtiss OX-5 engine of 90 dubious horsepower. He was a quiet, congenial young man who was popular with everyone in the company, and he had learned how to coax maximum performance from an airplane. [10]

After a series of engine runs on the ground to check operation of the oil, fuel, and ignition systems, on Saturday, August 18, 1929 Clark donned his parachute and leather flying helmet and marched onto Travel Air Field for a date with destiny. Until that day the fastest airplane he had flown was a 150-mph. D4000 with "speed" wings. Herb Rawdon's creation, however, was capable of at least 200 mph.

Nearby sat the Type R, resplendent in its bright red paint that was tastefully trimmed with black scallops and mint green accent striping around the nose and on the wings. The airplane had been painted by William "Frenchy" Hauselman. As Clark approached the racer a bevy of engineers and mechanics were swarming over the monoplane, making last-minute checks. The experimental crew had worked through the night to complete small details, and that morning they had finally pronounced the airplane ready to fly. Rawdon and Burnham, along with Walter Beech and a few other company officials, perused the machine one last time. In true test pilot form, Clarence made his own careful, methodical inspection and was satisfied that all was in proper order.

After a brief discussion with the engineers, Clark clambered up on the left wing root and slowly inserted himself into the cramped cockpit. He spent the next few minutes getting adjusted in the seat, then checked all controls for freedom and correct direction of movement. He next checked the fuel load. The 42-gallon fuel tank, which sat almost directly on the airplane's center of gravity, was located at the 24% mean aerodynamic chord point of the wing. A five-gallon reserve reservoir was installed directly aft of the main tank, but a wobble pump in the cockpit had to be used to transfer the fuel. Both tanks were made from .040-inch aluminum alloy, as was the engine firewall.

With mechanics and fire extinguishers standing-by, the propeller was pulled through several times to prime the engine. Clark retarded the spark control, positioned the magneto switch and shouted, "Contact!" A mechanic swung the propeller with all his might. The big Wright coughed once, then twice as puffs of black smoke were ejected from the exhaust stacks. The Type R shook as the nine-cylinder radial rumbled to life and settled into its arythmic, staccato idle. Flying wires trembled and the wings rocked up and down lazily. To Clark, the entire airplane seemed to quiver with anticipation.

The air swirling around the cockpit was saturated with the acrid smell of exhaust fumes as Clarence advanced the spark, checked the radial's vital signs, and nodded his head in approval as he gave Rawdon a thumbs-up. Gesturing to remove the chocks, he leaned to the left and advanced the throttle, struggling to see ahead

The third racer built for the 1929 National Air Races was the B11D powered by a seven-cylinder Wright J-6-7 engine rated at 240 hp. It was registered R612K and was the smallest biplane built by Travel Air. Flown by William H. Emery, it placed fourth in the Portland, Oregon to Cleveland, Ohio Derby that preceded the National Air Races. The aircraft later was bought by Douglas Davis who flew it for many years. ***(Raytheon Aircraft Company)***

as the airplane moved forward under its own power for the first time. After reaching the end of the runway he turned the ship into the wind and began a ritual he had performed hundreds of times before in his career. Holding the stick all the way back to keep the airplane's tail on the ground, Clarence ran up the engine to check the ignition system. The differential in RPM between the left and right magnetos was within limits. Next, flight controls were checked and then checked again. The position of the fuel valve was verified. Oil and cylinder head temperatures were scrutinized. Finally, there was nothing left to do but fly.

A small group of spectators had gathered along the runway to witness the great event. Taking another look at the instrument panel, Clark adjusted his goggles until they were snug against his face. He pulled the stick back and eased the throttle forward, swinging the tail around to line up with the 1/2 mile-long grass runway.

Clark made a final, quick scan of the instruments and slowly advanced the throttle to full power. The 420-hp. Wright roared in response as the little red racer accelerated like a bullet across the virgin sod, pushing the pilot back hard against the seat. The noise became deafening as the bellowing of nine exhaust stacks multiplied into a howling crescendo. Having traveled barely 200 feet down the runway, Clarence was playing a skillful dance on the rudder pedals to keep the ship traveling in a straight line as he cautiously eased the stick forward and lifted the tail into takeoff attitude.

When airspeed reached about 70 mph. Clark's skilled hands told him the wings were pregnant with lift. He eased the stick aft and the Type R took flight to the cheers of everyone on the sidelines. Leveling about 100 feet off the ground, Clarence held the nose down and let airspeed build to well over 100 mph. Suddenly, observers were horrified to see the aircraft descend sharply. They braced for the inevitable crash, but the red machine then began to climb steeply for altitude (Clark would later explain that he was only checking elevator control response. Nothing was wrong with the airplane).

Flying to the east of Travel Air Field, Clarence spent the next 20 minutes probing the airplane's response to pitch, roll, and yaw inputs. He found the Type R a delight to fly, and it made no special demands on the pilot. Stall speed was about 70 mph., which was benign for an airplane with a wing loading of 15.5 pounds per square foot. The airplane was stable in all three axes and could be flown hands-off "indefinitely," according to Rawdon. Cruising speed was 150 mph. at 1,550 RPM. Clark also found that the ship recovered rapidly from left or right spins by using standard recovery techniques or by neutralizing the controls.

After completing these maneuvers Clark attempted to determine the aircraft's maximum speed. He trimmed the aircraft and applied full power. The maximum indicated airspeed settled at 185 mph. He later reported to Rawdon that the airplane was tail heavy at that speed, but only if the stabilizer was trimmed to its upper limit. Rawdon attributed this characteristic to forward movement of the center of pressure on the wing at high speeds (tests conducted later with the cowling installed also revealed that in dives approaching 275-300 mph., stick forces again became heavy because of the shift in the center of pressure).

On the way back to the airfield Clarence knew Rawdon and Burnham would be pleased with the results of their hard work. Maintaining about 80 mph. on final approach, Clark put the racer into a steep sideslip so he could see the runway clearly. He brought the monoplane in for a gentle landing, taxied up to the factory, and shut down the panting Wright. When Rawdon walked up to the cockpit and saw Clarence wearing a big smile, he knew the maiden flight had gone well. Clark told him that the Type R was fast and that it was a "pilot's airplane" in every respect. There were a few minor adjustments to be made, but that was to be expected for a new airplane. These included reconfiguring exhaust stacks on the lower cylinders to expel gases out the bottom of the cowling instead of along the sides of the fuselage, which created too much heat in the cockpit. Clark recalled in 1981 that the Type R was "the best-flying Travel Air" he ever flew, and that it "had no tricks" and did not tend to ground loop during landing.

During the next two weeks the airplane made at least 12-14 test flights in preparation for the National Air Races. Both Herb and Walter were hoping that with the cowling installed maximum speed would increase to about 205 mph. They were wrong. Instead, it increased to more than 225 mph. In tests flown over a measured, two-mile course the average of two runs (one upwind, one downwind) was 227.5 mph. As the NACA tests had shown, the cowling provided a significant increase in performance. The cowl installed on the Type R had been modified slightly from the NACA design, and cooling baffles were not used between the cylinders. Oil and cylinder head temperatures, however, remained within limits.

The chief problem encountered during the flight test program was retaining the cowling in its mounts. As Rawdon later explained, "The aerodynamic bursting forces on the cowling were quite large and there was a large anti-drag component on the outer shell. Partial failures of the cowling occurred because of inadequate structural design, and during one flight the cowling was pulled forward into the propeller." Aerodynamic forces acting on the metal panels in the forward fuselage also were greater than anticipated, and some "beefing up" of the structure was required, according to Rawdon. In addition, he later concluded that although "controls are all adequate and effective, it would be advisable if building another airplane of this type to increase rudder area slightly" (the third, fourth and fifth airplanes built featured a larger vertical stabilizer and rudder).

Having completed an abbreviated flight test program, by the third week in August the Type R was declared ready for the Cleveland competition. It had taken Herb Rawdon and Walter Burnham 11 months to transform a concept scribbled on paper into a scarlet marvel. Although neither man could have realized it at the time, they had developed one of the earliest "X"-planes that would push the limits of aeronautical knowledge and set the pace for future designers to follow.

THE TRAVEL AIR COMPANY

"The Standard of Aircraft Comparison"

Formed late in 1924 by Walter H. Beech, Lloyd C. Stearman, Clyde V. Cessna, and Walter P. Innes, Jr., the Travel Air Manufacturing Company rose from humble beginnings in a small shop with a few hundred square feet of space near downtown Wichita, Kansas to five

Clarence Clark (right) poses with Truman (left) and Newman Wadlow beside a Curtiss-Wright Travel Air 6B cabin monoplane at Bartlesville, Oklahoma in 1983. Clark, who died in 1988 at age 84, was the only pilot to fly four of the five Type R monoplanes. In July 1931 Newman Wadlow performed flight tests on the last Type R built. He died in 1989 at age 82. ***(Phillips)***

The Travel Air factory complex as it appeared in the summer of 1929. Both R613K and R614K were constructed in Building "C" in the left foreground. **(Raytheon Aircraft Company)**

buildings on East Central Avenue that encompassed more than 166,000 square feet by August 1929. The company eventually produced more than 1,400 airplanes and employed nearly 700 people at the peak of its success.

Travel Air's first product was the Model A biplane that was designed by Lloyd Stearman and placed into limited production early in 1925. Powered by a Curtiss OX-5 engine, the biplane cost $3,500 but offered improvements in performance and efficiency that were lacking in aging, surplus World War One aircraft such as the Curtiss JN-4 and the Standard J-1. By 1930 more than 600 Model A and Model 2000 biplanes had been produced. In 1926 Stearman resigned to establish his own company in California. Cessna did the same in 1927 to satisfy his desire to build a full-cantilever monoplane, but he remained in Wichita.

Walter Beech became president of Travel Air and led the company until it was absorbed into the giant Curtiss-Wright Corporation in August 1929. After the stock market crash in October of that year the fortunes of Travel Air, along with the entire American aviation industry, quickly evaporated. Within months sales plunged and layoffs began. Production slowed to a trickle. Struggling to stay in business, Curtiss-Wright eventually was forced to close the Wichita factory. Seeking a new challenge, in April 1932 Walter Beech resigned from Curtiss-Wright and founded the Beech Aircraft Company.

One building of the idle Travel Air factory campus was rented by Clyde and Eldon Cessna in 1932 to build their diminutive CR-2 and CR-3 racers, and in April 1934 Walter Beech leased the entire complex from Curtiss-Wright to begin volume production of the legendary Beechcraft Model 17 biplane. Walter and his wife (and company co-founder) Olive Ann Beech eventually purchased the site and it became the permanent home of the Beech Aircraft Corporation.

Travel Air left a legacy that few aircraft manufacturers can claim. The company produced some of the most significant commercial airplanes of the 1920s, including the Model A, the Type 5000 cabin monoplane that became the first commercial airplane to fly from California to the Territory of Hawaii; the famous Type 5000 *Woolaroc*, winner of the Dole Race in August 1927, the Model 6000 cabin monoplane that set a new standard for business aviation, and the famous Type R racers. Travel Air was truly worthy of its slogan, *"The Standard of Aircraft Comparison."*

1. The quote is from aeronautical engineer Edward Swearingen and was made during an interview with the author in 1993.

2. Impressed by the popularity of the land-based James Gordon Bennett races that began in 1909 in France, Jacques Schneider sponsored a similar competition for seaplanes in an effort to stimulate commercial development of these aircraft. The first race was held in April 1913 at Monaco. There was no competition during World War One, and the races resumed in 1919 and eventually were held on an every-other-year schedule. The British won the trophy for the third time in 1931 and the races were terminated. Although the Schneider events did little to spur commercial development of seaplanes, they exerted a lasting influence on air racing and aircraft design in the years that followed.

3. Different references to the airplane were used by Rawdon. In addition to "Type R racing airplane," he also used "RB" which stood for "Rawdon/Burnham." The term "R-100" and "RA-100" also appear in documents. There was no official designation used by the Travel Air Company, chiefly because the design did not receive an Approved Type Certificate and was not placed into production.

4. Prior to 1927 there was no official body of regulations governing aviation. As the industry grew during the mid-1920s, it became obvious that federal regulation was necessary to manage growth and improve safety. Government and industry worked together closely to fashion the new rules governing flying and airplane design and construction. The handbook was the result of input from many engineers and designers, and made an important contribution to the science of aeronautics.

5. In February 1999 Dave Jameson told the author he did not know the origin of the short-span set of wings donated to the Staggerwing Museum Foundation, and that he had only assumed they were the set built for R614K. Jameson said he had no documentation or other information that would serve as evidence to support that contention. In addition, he agreed that the wings could have been built specifically for *Texaco 13*.

6. The M-6 airfoil was developed by the National Advisory Committee for Aeronautics, the forerunner of the National Aeronautics and Space Administration. Hundreds of airfoil designs were available in the late 1920s. Choosing the correct section for the airplane was a major decision. Rawdon had experience with the M-6, but the RAF 34 proved superior for use on a racing airplane because of its more benign stall characteristics.

7. After years of research and inquiries by the author, the identity of the engine designer remains unknown. There were a number of people in Wichita at that time who were trying to develop engines, but few had the resources to succeed. There is no evidence, however, that the visitor to Travel Air was a local resident. According to Rawdon, the engine was not developed and the proposal was quickly dismissed by Beech.

8. Frank Hawks and Gerald Vultee were destined for success and notoriety in the aviation industry. Hawks eventually became a well-known pilot who set many speed records, and Vultee founded his own airplane company and was a respected designer. The Lockheed *"Air Express"* was powered by a Pratt & Whitney *"Wasp"* radial engine. It could carry passengers as well as mail and cargo. The record flight across America electrified the public and thrust Hawks into the limelight for the first time.

9. Sunyak, John: "The Mystery Ship Mystery." *Skyways* magazine, No. 37, January, 1996, pages 68-70.

10. Clark was a key employee of Travel Air. Between 1926 and his departure from the company in 1930, he test-flew hundreds of new airplanes in preparation for delivery to their owners. Clarence reported directly to Walter Beech, not to factory manager William Snook. Clark flew four of the five Type R airplanes built and was considered an expert aviator by his contemporaries including Charles Lindbergh, whom he knew well. The author interviewed Mr. Clark several times before his death in 1988.

The Men Behind the Type R

CHAPTER TWO

Herbert M. Rawdon

Born in Kansas City, Missouri in 1904, Herb Rawdon grew up in Woodward and Sharon, Oklahoma. During high school he worked during the evenings and on weekends at an ice cream factory, but he also was hired to work on a farm and a ranch during summer vacation each year. After graduating in 1921 Herb became a full-time employee of the ice cream company for the next year.

He had an interest in engineering and enrolled initially at Ohio Northern University before transferring to Tri-State College in Angola, Indiana. To earn extra money he tutored students in mathematics and mechanics, and worked in a restaurant. Rawdon gleaned some engineering experience one summer when he worked for the Goddard Construction Company in Detroit, Michigan. His job was supervising six men that were laying down steel reinforcements for buildings. He also found work at a nearby aircraft factory and taught engineering drawing at Tri-State College in classes attended by more than 100 students.

Rawdon earned a Bachelor of Science degree in mechanical engineering in 1925, and soon found work as a draftsman for the Fuller-Leheigh Engineering Corporation in Pennsylvania. His work centered on making design drawings of industrial powerplants. In 1926 he joined the fledgling Travel Air Manufacturing Company in Wichita, Kansas. He spent the first year working as a factory mechanic building, assembling, and rigging biplanes. According to historian Bruce Bissonette, in 1926 Rawdon also assisted Clyde Cessna with the design and construction of a semi-cantilever wing installed on Cessna's Anzani-powered cabin monoplane, which served as the basis for the Travel Air Type 5000 transport introduced in 1927.

Rawdon's engineering talents eventually were recognized by Walter Beech, and in his second year with the company Herb was promoted off the shop floor and into the engineering department where he gradually worked his way up to assistant engineer. By 1930 he had been promoted to chief engineer, supervising 25 men and a shop force of 20 men. He was responsible for all engineering work as well as research and development projects (such as the Type R), and studied aircraft design theory in his spare time. Predicated on his experience with NACA-type cowlings and Townend rings, Rawdon wrote a technical paper entitled *"Cowling Radial Aircraft Engines To Reduce Drag."* It was presented before a local group of the National Society of Automotive Engineers, but the paper remained in unfinished form and was not published.

Herbert Rawdon was a 25 year-old engineer with the Travel Air Company when he conceived the design that would become the Type R monoplane. ***(Raytheon Aircraft Company)***

Following Travel Air's absorption into the giant Curtiss-Wright Corporation in 1929, Rawdon remained with the company until the Wichita facilities were closed in 1932. Although he was given an opportunity to transfer to another division at the same salary and with similar responsibilities, he declined "on account of unfavorable internal affairs which did not seem sufficiently promising to warrant the transfer," Herb later wrote. [1]

He formed the Rawdon-Christopher Aircraft Company that specialized in maintenance and modifications, but after that business failed he worked as a draftsman for Lockheed and Boeing during 1933-1934. In 1935 he served as an engineering instructor for the

Curtiss-Wright Technical Institute in California before being hired as a production manager and superintendent for the Spartan Aircraft Company in Tulsa, Oklahoma. From 1937-1940 Rawdon worked as a design engineer for the Douglas Aircraft Company in El Segundo, California as well as acting as a consultant for the National Aircraft Company in San Antonio, Texas.

With war clouds gathering on the horizon in 1940, he returned to Wichita and went to work for his old boss Walter Beech at the Beech Aircraft Corporation. He served as chief engineer for the company's Design and Research Division until resigning in 1960. Remaining independent to work on advanced aircraft designs of his own, Rawdon served as a consultant for Boeing's Wichita Division, the Cessna Aircraft Company, and the Lycoming engine company in Williamsport, Pennsylvania.

In partnership with his brother Gene Rawdon he later formed the Rawdon Brothers Aircraft company located within 1/2-mile of the old Travel Air factory on East Central Avenue. In 1938 he designed the Rawdon R-1, a low-wing monoplane intended to compete in the pre-war Civilian Pilot Training Program, but it was rejected by the government. He later designed the Rawdon T-1 sport trainer that was similar to the R-1 but was powered by 125-hp. Lycoming engine. The airplane was certified in 1947 and could be equipped to spray crops. The brother's most successful product, however, was the "Rawdon Hatch"—a cockpit enclosure for war-surplus trainers such as the Fairchild PT-19 and PT-23. About 800 were sold.

Herb Rawdon held a private pilot license and was an accomplished aviator. After a distinguished career in aviation that spanned nearly 50 years, he died on December 2, 1975.

Walter E. Burnham

Seven years older than Herb Rawdon, Walter Burnham was born in Knobnoster, Missouri in September 1897. He received his elementary school education in Horton, Kansas before his family moved to Wichita when he was in the eighth grade. He graduated from high school in 1916.

After his father died Walter took a job with the Martin Metal Manufacturing Company as an apprentice. He enjoyed the work and had been promoted to an assistant manager before leaving to enlist in the United States Army's Student Army Training Corps in 1918. He was assigned to the University of Kansas at Lawrence where he entered the school of mechanical engineering. Unfortunately, during the final three months of the war he was placed on active duty and was unable to graduate.

After his discharge from the military Burnham returned to Wichita and resumed his old job with the Martin Metal Manufacturing Company. He continued studying engineering through extension programs. In the mid-1920s he joined the Travel Air Manufacturing Company where he served as an engineer until the aviation market collapsed and the factory was closed in 1932. That year he helped design and build a wind tunnel at Tri-State College in Angola, Indiana and while there earned his degree in aeronautical engineering. Theodore "Ted" Cochran, a co-worker on the Type R project at Travel Air in 1929, recalled that Burnham was "the best all-around aircraft mechanic, woodworker, welder, sheet metal man and engineer I ever met."

When the wind tunnel project was completed, Burnham joined the Tri-State faculty as a laboratory instructor and in 1934 was appointed head of the aeronautical engineering department. He held that position until 1937. That year he and Gordon L. Freedman established the Freedman-Burnham Engineering Corporation in Cincinnati, Ohio. The company specialized in the design and construction of propellers for light aircraft and had soon developed a reputation for quality products.

Walter remained with the company until 1944 when he was hired by the Beech Aircraft Corporation to work as a research engineer alongside his old friend, Herb Rawdon. During this time, however, Burnham continued to do consulting work for the Freedman-Burnham company. He resigned his job at Beech Aircraft in 1949 and joined the Boeing Wichita Division where he remained until his retirement in 1965. He was chief tooling engineer before his reassignment to manager of the manufacturing, research and development department, which he supervised until retiring.

In addition to his work in the aviation industry Walter served as a consulting engineer for a number of businesses including the McCauley Industrial Corporation and the Highway Aircraft Company. He designed a propeller for McCauley that was intended for use on a target

Walter Burnham was known at the Travel Air factory as a highly competent engineer with talents for welding as well as wood and sheet metal work. Rawdon's choice of Burnham for a partner in the Type R project was fortuitous. ***(Raytheon Aircraft Company)***

drone developed by Radio Plane, and also designed a propeller to be used on an automobile invented by the Highway Aircraft Company. The car was a failure, and for years afterward Burnham became entangled in a legal struggle with the company.

Burnham played the viola and was active in the Friends University Community Orchestra. After retiring he continued to do consulting work, but was able to spend much of his time building and repairing different types of stringed instruments. He was involved with the York Rite, was an elder in the United Presbyterian Church, and served on the board of the Civic Music Society of Wichita. Walter Burnham died on February 7, 1976.

Walter H. Beech

Walter H. Beech, photographed in 1929 at the time of the Travel Air's acquisition by the Curtiss-Wright Corporation. His zeal for air racing helped ignite Herb Rawdon's plans for building a high-performance racing monoplane. ***(Raytheon Aircraft Company)***

An aviation legend in his own time, Walter Beech was born in January 1891 and was raised on a farm near Pulaski, Tennessee. Unlike Herb Rawdon and Walter Burnham, Beech lacked a formal education but he possessed a keen insight and understanding of farm machinery and mechanical systems. A voracious reader, he increased his knowledge and vocabulary by studying dictionaries.

In 1911 he became a chauffeur in Minneapolis, Minnesota and worked as an automobile mechanic learning valuable lessons about the design and operation of reciprocating engines. During World War One Beech enlisted in the United States Army and became a competent and respected aviation engine technician at the Army's Signal Corps Aviation School at Rich Field, Waco, Texas. Contrary to myth, he was not a flight instructor during the war but did learn to fly in 1919 as an enlisted pilot. He flew a Curtiss JN-4D biplane for the majority of his training.

He was discharged from the Army in 1920 and was hired as a pilot for the Williams-Hill Airplane Company in Arkansas City, Kansas before relocating to Wichita in 1921 where he worked for the Wichita Laird Airplane Corporation operated by Emil "Matty" Laird. By 1923 Laird had moved to Chicago and Beech was general manager of the renamed Swallow Airplane Manufacturing Company under the direction of Jacob Melvin Moellendick. Walter, along with Lloyd Stearman, William Snook, and Clyde Cessna formed the Travel Air Manufacturing Company late in 1924 and built a three-place, open cockpit biplane powered by a 90-hp. Curtiss OX-5 engine. Designed chiefly by Stearman, the Model "A" was a success and the business and Travel Air's product line grew rapidly. By 1928 Walter Beech had become president following the departures of Stearman and Cessna to form their own companies.

Under his leadership Travel Air became one of the world's premier airplane manufacturers and was a major builder of commercial aircraft in the United States. Beech led the company through its merger with the giant Curtiss-Wright Corporation and in 1930 he was placed in charge of the Curtiss-Wright Sales Corporation. The Great Depression forced the company to close the Travel Air factory in 1932. In April of that year, however, Walter and his wife Olive Ann, along with engineer Ted Wells, formed the Beech Aircraft Company in Wichita to

The engineering department at Travel Air in the autumn of 1928 when Herb Rawdon and Walter Burnham began preliminary design work on the Type R monoplane. Horace Weihmiller is sitting at his desk. Rawdon is in the rear of the room at the second table. Engineer and draftsman J. Jack Clark, who helped design a number of Travel Air monoplanes, is sitting in front of the door. ***(Raytheon Aircraft Company)***

Walter Burnham and Herb Rawdon (second and third from left, respectively) pose for the camera with pilot Arthur Goebel (seventh from left) and the Type 5000 monoplane "Woolaroc" that Goebel and navigator William Davis flew from California to the Territory of Hawaii in August 1927 to win the Dole Race. The men shown also built the two Type R machines in 1929. They include Howard Bacchus (far left), William Hauselman (fourth from left), "Pinky" Grimes, Theodore "Ted" Cochran, (Goebel), Ralph Morton, Harold Brooks, and Clarence Clark (far right). Engineer Horace Weihmiller (next to Clark) designed the Type 6000 cabin monoplane in 1928. The Woolaroc eventually was restored and placed on permanent display at the Frank Phillips museum in Bartlesville, Oklahoma. ***(Raytheon Aircraft Company)***

An informal photograph of the Travel Air engineering force in 1929. Herb Rawdon is kneeling in front, with Walter Burnham in the front row, second from right. Theodore "Ted" Wells, an apprentice engineer who in 1931 would design the Beechcraft Model 17 cabin biplane, is standing in the second row, sixth from left. ***(Raytheon Aircraft Company)***

build and sell the Model 17 biplane. In 1934 Beech relocated the business to the vacant Travel Air facilities on East Central Avenue. The company received huge orders for multi-engine training airplanes during World War Two and prospered in the vibrant, post-war market for business aircraft.

In 1945 Beech strongly advocated the design and construction of a new, four-place airplane that could carry people and baggage at speeds up to 170 mph. The result was the Beechcraft Model 35 *"Bonanza"* that remained in production for 35 years. Walter H. Beech died in November 1950 after a distinguished, 31-year career in aviation. He was inducted into the Aviation Hall of Fame in 1977 and the Kansas Hall of Fame in 1987. His wife Olive Ann Beech, who became a legend in her own right, died in July 1993. She was inducted into the Aviation Hall of Fame in July 1981.

Cracking a rare grin, Beech poses beside the Travel Air "Special" that he flew to victory in a free-for-all race at Tulsa, Oklahoma in August 1925. Beech eventually logged more than 10,000 hours in the air before encephalitis forced him to stop flying by 1942. ***(Raytheon Aircraft Company)***

CHAPTER TWO NOTES

1. Rawdon probably was concerned that the transfer to Curtiss-Wright's St. Louis factory would result in his being dismissed after a short time. He preferred to remain in Wichita to find other employment. The only Travel Air engineer that accepted a transfer was Theodore "Ted" Wells. At St. Louis he served as a production test pilot as well as an engineer. In 1932 Wells joined with Walter and Olive Ann Beech to establish the Beech Aircraft Company in Wichita. Ted's design for negative stagger-wing biplane would become the Beechcraft Model 17R.

In 1926 Beech and navigator Brice Goldsborough (right) won the prestigious Ford Reliability Tour in a Travel Air Type BW equipped with special Pioneer instruments. Note the compass installed beneath wing center section and drift meter attached to the aft cockpit. Recessed exhaust manifold was developed by Travel Air engineers. ***(Raytheon Aircraft Company)***

In May 1929 the Travel Air office staff and engineers were photographed dressed in their business attire. Four key individuals are visible: Herb Rawdon (back row, fourth from left), Olive Ann Mellor (back row, seventh from left), Walter Burnham (kneeling at far right), and Merle Van Buskirk (back row, third from right) who accompanied Walter Beech to New York as his secretary after the company was absorbed into the Curtiss-Wright Corporation. Olive Ann Mellor and Mr. Beech were married in 1930. Rawdon's wife, Laura Lee, is in the second row at far left. ***(Raytheon Aircraft Company).***

Walter Beech learned to fly in 1919 at Rich Field near Waco, Texas. He is standing second from the left in the photograph. The airplane is a Curtiss JN-4 "Jenny" powered by a 90-hp. Curtiss OX-5 engine. Sergeant Beech became an expert engine mechanic during his service at Rich Field before becoming an enlisted aviator. ***(Walter H. and Olive Ann Beech Collection, Ablah Library, Wichita State University)***

Walter Beech (left) poses beside a Laird Swallow early in the 1920s. E.M. Laird, designer of the airplane, is third from left. ***(Special Collections, Ablah Library, Wichita State University)***

A Marvel of the Age

R-2001 - (N)R614K

Late in August 1929 the Travel Air factory was buzzing with activity as a small fleet of airplanes and a group of pilots prepared to depart for the National Air Races in Cleveland. It was an exciting time as mechanics made last-minute inspections of each aircraft, filled fuel and oil tanks, and the planned route of flight was reviewed by each pilot. In addition to the Wright-powered Type R (R614K) and her Chevolair-powered sister ship (R613K), a Model 6000B and one of the new, four-place Model 10 cabin monoplanes would accompany the racers on their aerial trek north. Among the passengers in the larger monoplane were Walter Beech and Herb Rawdon, and company pilot Newman Wadlow would fly the Model 10 accompanied by passengers that included Theodore "Ted" Cochran, a Travel Air mechanic. Both of the cabin ships were scheduled to be placed on display at the Cleveland aircraft exposition to be held concurrently with the races.

Anxious to be on his way, Clarence Clark nestled into the cockpit of R613K and prepared to start the in-line, six-cylinder engine. He had hoped to fly the Wright-powered racer to Cleveland, but that privilege had been reserved for a Travel Air dealer from Atlanta, Georgia who also was a long-time friend of Walter Beech. His name was Douglas Davis. Beech had chosen Davis over Clark chiefly because he had more experience in pylon racing and to provide Davis with publicity for his Travel Air dealership. Clarence was disappointed, but he accepted the decision without complaint.

Davis had arrived at the factory only a few days before and was introduced to the mighty Type R by Clark, who briefed Davis on the airplane's handling characteristics. Doug had made two or three flights and totaling about 1 1/2-hours in the red and black speedster and felt comfortable with the machine. Although it was the fastest airplane Davis had flown and demanded more skill than the biplanes he was accustomed to flying, he was pleased to have been chosen to pilot the aircraft and his endless enthusiasm for the ship seemed to infect everyone, even Walter Beech.

But Doug's unfamiliarity with the high-performance Type R did result in a scare. Both Clark and Rawdon later recalled that when Davis took off for Cleveland he made a diving, low pass over Travel Air Field at an estimated 250-275 mph. As the airplane flashed by a small crowd that had gathered to witness the departure, the landing gear struck the ground—not hard but at a high

Photographed at the National Air Races in Cleveland, Ohio R614K rests in the grass next to a Great Lakes biplane. Doug Davis performed aerobatics in the Travel Air that stunned the crowds as well as other pilots. The Type R's power-to-weight ratio gave it impressive vertical capability. Note the different shade of the letter "R" on the rudder compared with "614K." ***(Raytheon Aircraft Company)***

Orthochromatic film provides almost no contrast between the red and black colors on R614K, photographed as it taxied for takeoff. ***(Peter M. Bowers via John W. Underwood)***

speed. Sensing a firm thud, Davis immediately pulled up and gained altitude to assess the situation. Nothing seemed to be wrong so he continued on toward his first fuel stop at Kansas City. During the landing, however, Doug knew that something was amiss. He quickly realized that both main wheel tires were blown. Mechanics installed two new wheels and tires that were obtained locally and Doug continued his flight northward.

At Cleveland Walter Beech had made preparations to house both Type Rs in a hangar on one side of the airfield to keep them out of sight until after the races had begun. Both R614K and R613K were quickly rolled into the hangar and covered to hide their forms from curious spectators. News of the mystery ships' arrival spread quickly and during the next few days Beech skillfully worked the press, feeding them tantalizing bits of information but never revealing too many key details about the airplanes. No one was allowed in the hangar except Travel Air employees and members of the racing team, according to Theodore "Ted" Cochran and Newman Wadlow. [1]

The United States Army Air Corps entered a modified Curtiss "Hawk" in the Free-for-All race, Event No. 26 at the Cleveland races. Designated XP-3A, the biplane was powered by a Pratt & Whitney "Wasp Jr." radial engine surrounded by a NACA-type cowling. Bayonet-type exhaust stacks were intended to generate a small amount of thrust. A series of shutters (visible behind the propeller spinner) could be controlled from the cockpit to adjust engine cooling. ***(Raytheon Aircraft Company)***

The crowds were thick at the 1929 National Air Races held from August 24 through September 2. With more than 40 races scheduled during the show, thousands of people jammed into the grandstands at Cleveland Municipal Airport to see the best of America's airplanes and pilots battle for greenbacks and glory. They would not be disappointed. John T. Nevill, an experienced and knowledgeable reporter for *"Aviation"* magazine, determined that an average of 417 airplanes were on the airport grounds each day.

Pilots from the United States Army Air Corps and United States Navy teams, which Beech suspected would be the only serious competition for the Type R, garnered less attention from reporters than they were accustomed to receiving. Their two airplanes essentially were modified fighters with more powerful engines and streamlining to reduce drag. Although technologically inferior to the Type R, both airplanes were worthy adversaries.

This was especially true of the Army's entrant, a Curtiss XP-3A *Hawk*. It was the second XP-3A built by Curtiss (c/n 28-189) and had been used in development of the NACA cowling. As did the Type R, the Army biplane featured a NACA cowl but featured a large propeller spinner. Bayonet-type exhaust stacks extended aft of the cowling, and bulbous fairings were installed along the fuselage sides to minimize interference drag between

The United States Army Air Corps XP-3A featured large, bulbous fairings along each side of the fuselage to reduce interference drag between the cowling and the fuselage, and the pilot's windshield was modified to improve airflow around the cockpit. The handsome racer was flown by Captain R.G. Breene. ***(Raytheon Aircraft Company)***

After its victory in the Thompson Cup contest R614K awaits crowds of admirers. The forward fuselage is streaked with oil and exhaust stains and the soft aluminum wheel fairings have been dented by stones blown back by the propeller during taxi and takeoff operations. **(Kenneth D. Wilson)**

A pair of military pilots inspect R614K. The cumbersome, two-piece cowling was later abandoned in favor of a four-piece design installed on the Shell and Texaco Type R machines in 1930. **(Peter M. Bowers)**

the narrow fuselage and the round engine cowl. The biplane was powered by a Pratt & Whitney R-1340 radial engine rated at 450 hp. and had an empty weight of 1,955 pounds. The pilot was Captain R.G. Breene. After the 1929 races it was converted into the second XP-21 and later became a P-1F pursuit ship. [2]

As for the Navy's entrant, it was a Curtiss F6C-6 *Hawk* (Navy A-7144) that was powered by a 12-cylinder, Prestone glycol-cooled Curtiss *Conqueror* D-12 engine developing 435 hp. The landing gear and airframe received minor streamlining to reduce drag, but not to the extent of the Army's XP-3A. The pilot was Lt. Commander J.J. Clark. After the 1929 races it underwent major modifications into a parasol-wing monoplane designated the XF6C-6 and was entered in the 1930 Thompson Trophy race.

Although Beech could have entered R614K in a number of preliminary races, he kept the airplane out of the public eye until Doug Davis took it aloft and performed a mild but impressive aerobatic routine. He flew a series of loops and aileron rolls as well as performing vertical rolls. An observer at the races noted that "The machine had such an enormous reserve of power that it seemed to travel in any attitude or direction, including upside down flight and flights vertically upward, quite normally and under full control."

Doug Davis beams with pride as he holds the Thompson Cup after winning Event No. 26 at the Cleveland races. Beginning in 1930 the speed contest would become the famous Thompson Trophy Race that dominated the National Air Races throughout the 1930s. **(Kenneth D. Wilson)**

A well-dressed Doug Davis poses with the sanitized R614K. The airplane created a sensation and sent the military back to the drawing board to reconsider the merits of monoplanes. **(Kenneth D. Wilson)**

Except for this brief exhibition, Walter Beech was preserving his low-wing warrior for the one race it had been designed and built to win—the free-for-all event scheduled for September 2. It was to be the highlight of the week-long aerial extravaganza, and as the days went

Newman Wadlow (left) and a Travel Air mechanic strike a casual pose with R614K at the Cleveland races. ***(Raytheon Aircraft Company)***

by speculation arose about whether the racer from Wichita was capable of defeating the best the military had to offer.

When race day dawned, Herb Rawdon and his team were ready. They had been busy since early in the morning preparing R614K for the upcoming contest. A representative from the Wright company checked the engine and magnetos thoroughly and installed new BG spark plugs, and the cowling was checked for proper fit and retention. Wearing race number 31 on both sides of the fuselage and the underside of the right wing, the racer was cleaned and buffed to a brilliant shine. The propeller blade angle was decreased to allow the engine to turn a maximum of 2,700 RPM., and the fuel and oil tanks were filled to proper levels for the race. All R614K needed was her pilot.

Doug Davis was rested, ready and eager to fly that day. Earlier in the week he had won the relay race for civilians in an OX-5-powered Travel Air biplane, and he had flown the Chevolair-powered Type R (race number 32) to victory in the experimental ship race. He had won an impressive $1,150 for his efforts. But unlike the previous races, Event No. 26 would push both man and machine to the limit in their quest to win $750 and a new award known as the Thompson Cup. The 1929 race would be the first contest to offer the prize, which was sponsored by president Charles E. Thompson of Thompson Products, Inc. [3]

As race time approached, Walter Beech gave Davis a few quick reminders and shook his hand as R614K was rolled out of the hangar and the powerful Wright J-6-9 rumbled to life. Doug taxied his steed across the airfield and was directed into position by officials at the starting line in front of the grandstands. Six other entrants in the free-for-all race also were there, including Captain Breene in command of the attractive XP-3A and Lt. Commander Clark in charge of the handsome F6C-6. The only other serious contender was Roscoe Turner in his Lockheed Vega powered by a 450-hp. Pratt & Whitney *Wasp Jr.* radial engine. The race consisted of 10 laps around a 5-mile, triangular course and the winner was determined by elapsed time, not speed.

It is interesting to note that the Army and Navy's *Hawks*, the Type R, and Turner's Lockheed were reasonably well matched in terms of horsepower, ranging from 420 hp. in the Type R to 450 hp. in Turner's Vega and Breene's Hawk. But at 1,950 pounds the Travel Air not only weighed less than its opponents, but had the advantage of being designed specifically for racing. It was clear to Beech and Rawdon, as well as other experienced pilots, that the contest probably would evolve into a duel between Davis and Breene.

Doug sat snug in the cockpit, occasionally exchanging glances with Breene and Clark in their trim biplanes. The officials walked up and down the line, making their final checks and ensuring that each pilot was ready. Walter Beech and Herb Rawdon watched anxiously from the sidelines, anticipating the start. Satisfied that all was in proper order, the starter stood in sight of all the contestants and raised his flag.

After a brief pause that seemed like an eternity to the fliers, the flag suddenly dropped. Breene took off first, followed at 10-second intervals by Davis, Clark, Turner and the others. The ground trembled as engines accelerated to full power and propellers blew billowing clouds of dirt and dust in their wakes as each aircraft surged forward, its wings searching frantically for lift. The race was on!

Crowds cheered and stood on their feet as the racers rounded the first pylon. Captain Breene had jumped into the lead, followed closely by Commander Clark and Davis. But Breene held the lead only briefly. The Type R soon passed the *Hawk*. As the race slowly unfolded Breene overtook three aircraft in the first few laps, and Davis had left two competitors behind as his machine continued to stretch its lead over the Army and Navy biplanes.

Both Breene and Davis flew their machines around each pylon as tightly as they dared, making the turns at an altitude of about 250-300 feet in about four seconds. Breene, however, dived about 60 feet in his turns, but Davis deviated little from his entry altitude. The *Hawk* was fast, but the Type R never relinquished its lead. As

Another candid photograph of R614K made after the Thompson Cup race at Cleveland. The small access door for the fuel tank is visible above the forward support strut. ***(Smithsonian Institution negative number 75-6488)***

Doug Davis talks with Travel Air mechanics before taxiing R614K for takeoff at the 1929 National Air Races. **(John Sunyak)**

the race progressed the monoplane slowly pulled away from the military ships and was setting a brisk pace. As the race neared its end, Doug feared that he had cut inside of a pylon—a violation that would disqualify him from the race. With his heart pounding like a jackhammer, Davis whipped the Travel Air around and flew back to circle the marker again. In his haste to complete the turn, however, he pulled back on the stick too hard and the high "G"-forces caused him to momentarily lose his vision. Meanwhile, Breene was closing the gap. Uncertain of whether he had successfully negotiated the turn, Davis quickly recovered and flew around the pylon one more time before rolling wings level and streaking down the straightaway.

In November he recalled the incident during an interview with *"The Atlanta Journal"* newspaper: "You know there is a limit to what a man can endure when he is flying at high speed. When you pass the 200 miles-per-hour mark you have to be careful on your turns. If you change direction too abruptly the blood rushes from your head and you go blind." He told reporters, "That happened to me in the Cleveland race. Near the end of the race I thought I clipped a pylon too closely. If I had failed to go around it my entry would have been thrown out of the meet. I had a long lead over the field, so I figured it would be better to circle the marker again and make sure I was around it. But this time I pulled up on the stick too quickly, and everything went black. By the time my head cleared I was already past the pylon, so I circled back once more. This time I took a larger curve and made sure I was outside the pylon. The judges told me after the race I had made a good turn each time."

In the remaining laps Davis managed to stay ahead of Breene in the pursuing *Hawk* and was first to cross the finish line after flying the race in 14 minutes 5 seconds. His fastest lap was 208.69 mph. and average speed was 194.9 mph. Captain Breene finished second with a time of 14 minutes 42 seconds and averaged 186.84 mph. Turner took third at 163.44 mph., and Commander Clark was fourth. In addition to winning, Davis and the Type R had made history by becoming the first civilian pilot and commercial airplane to defeat both the Army and Navy in unrestricted competition.[4]

In fewer than 15 minutes Travel Air's so-called *"Mystery Ship"* had turned the world of air racing upside down. Russell C. Johns, a journalist for *Aero Digest* magazine, wrote that the monoplane was "undoubtedly the most outstanding development among racing planes," and that its victory "focuses attention on the whole matter of America's status in pursuit aviation, and matters

Number "31" at rest in front of Building "D" at the Travel Air factory. **(Ben Bitler)**

View of R614K parked in front of the grandstands. The oil cooler is visible in the left wing stub. **(Kenneth D. Wilson)**

The Travel Air Company flew a six-place cabin monoplane similar to the Type 6B illustrated to Cleveland. It transported spare parts along with Walter Beech and Herb Rawdon, and was later displayed at the aircraft exhibition held concurrently with the air races. ***(Raytheon Aircraft Company)***

pertaining to radical changes in design which might help the Army and Navy keep up with progress." In the wake of Event No. 26, military officials began to seriously reassess their thinking about the monoplane and its inherent advantages. It is interesting to note that one year later the Navy would return to Cleveland with a monoplane of their own—the 700-hp. XF6C-6 that was built from the same F6C-3 that had been flown by Lt. Commander Clark.

Herb Rawdon was thrilled that the machine he and Walter Burnham had worked so long to create had proven its superiority over the military pursuits. His design concepts had been vindicated. Walter Beech, of course, also was elated at Davis's triumph. Rumors circulated around the airfield that no sooner had Doug landed and parked the airplane in front of the grandstands than Beech set about collecting on wagers he had made before the race. Theodore "Ted" Cochran recalled that "Walter Beech was going around the field taking up everybody's bet. He had a wad of bills that would choke a mule and he was really happy. After the races he cashed in and I imagine he probably made in excess of $60,000 on bets." It is improbable that Beech collected that amount of money, but he did hand out $100 bills to Cochran whenever he needed expense money.

As for Doug Davis, he climbed out of the cockpit and smiled broadly as he accepted the Thompson Cup amid a thunderous ovation from the crowd. True to his character, Davis had flown the race dressed in a business shirt and a striped tie. But he was not the only Travel Air pilot that achieved fame at the races that week. Aviatrix Louise Thaden had flown her Travel Air D4000 speedwing from Santa Monica to Cleveland in 20 hours, two minutes, two seconds to win the first Women's Air Derby (later known as the *"Powder Puff Derby"*). Her average speed for the grueling event was 135.97 mph., and she earned $3,600 in prize money as well as the *"Aerol"* trophy sponsored by the Cleveland Pneumatic Tool Company. Ted Wells had won the Portland, Oregon to Cleveland race in his trusty D4000, and William

A second cabin monoplane, the new Type 10 four-place design, also was displayed at the Cleveland exhibition. It was offered with a 225-hp. Wright J-6-7 or a 300-hp. J-6-9 radial engine. The Great Depression limited sales to only 10 or 11 airplanes. ***(Raytheon Aircraft Company)***

Emery had placed third in the diminutive B11D biplane that bore race number 30. The *"Rim of Ohio Derby"* was won by J.O Donaldson in a 300-hp. Travel Air biplane, and George Shealy and Herbert St. Martin were victorious in the civilian OX-5 race and the Canadian Commercial race, respectively. The 1929 National Air Races had been an outstanding success for the Travel Air Company.

With the Cleveland competition now history, the Travel Air team returned to Wichita and a warm welcome from employees and the public. In the autumn of 1929 the company was riding the crest of its popularity and orders were pouring in as customers clamored for OX-5 and radial-engine biplanes and the stately Model 6000-series cabin monoplanes. At the factory more than 600 workers in two shifts were laboring in a vain attempt to keep pace with demand as the rate of production approached 50 airplanes per week. [5]

Unfortunately, catastrophic events on Wall Street would soon clip the company's wings and plunge the aviation industry into a tailspin known as the Great Depression.

But those bleak days were still months away and few people at Travel Air feared for the future. That included the new Curtiss-Wright Corporation that had absorbed the Wichita manufacturer into its fold. Anxious to capitalize on the Type R's sudden notoriety, the racer was brought into the shop and refurbished in preparation for more racing and public appearances at local and regional across the United States. A restricted license for air racing had been issued by the Department of Commerce that was scheduled to expire on December 1, 1929. In addition, on October 9 the airplane's paperwork was changed to show the Travel Air Company as owner (on August 19 Doug Davis had been listed as the original owner after R614K was completed).

As the news spread about the airplane's success at Cleveland, the public relations department at Travel Air was kept busy responding to a flood of requests for information about the Type R from the press and would-be buyers. Although Doug Davis would fly R614K again, the airplane was placed under the command of Clarence Clark and Wichita pilot Ira McConnaughey, who flew for Universal Air Lines and was a seasoned and competent aviator.

Throughout September and October the two pilots made local and regional headlines with the airplane. After having its coat of red and black paint cleaned, polished and buffed to lustrous shine, the Type R was flown to the Kansas City International Exposition accompanied by its sister ship R613K, which had been re-engined with a 225 hp. Wright J-6-7 radial engine following its return from Cleveland. Piloting R613K Clark won the race for airplanes with engines having less than 800-cubic inch displacement, and McConnaughey and R614K took first place in the aerobatic contest and also won the free-for-all race with Clark finishing second in R613K.

The pilots next changed airplanes and Clark was victorious in another free-for-all event and McConnaughey

Walter Beech and Louise Thaden were caught on film at the Cleveland races. In 1927 Beech had helped Thaden get her start in aviation when she went to work in California for Travel Air dealer D.C. Warren. ***(Raytheon Aircraft Company)***

was second across the finish line. After these impressive performances by the two aviators the contest committee asked McConnaughey to fly R614K over a measured, five-mile course and attempt to set an official record for that distance. He agreed and flew the ship to a speed of 236.84 mph.—an unofficial record.

Clark also raced R614K at Sioux Falls, South Dakota at the Sioux Falls AirFair where he won the closed-course race beating Johnny Livingston in his taperwing Waco. Clarence also made "five short flights" at Sioux Falls, according to one of his logbooks pertaining to the autumn of 1929. Hot on the heels of the South Dakota competition, in October both racers were flown to Tulsa, Oklahoma to participate in air race events held in associ-

Louise Thaden won the Women's Air Derby from Santa Monica, California to Cleveland in a Travel Air D4000 powered by a Wright J-5 radial engine. Bearing race number "4" the airplane had a NACA-type cowling and streamlined landing gear. ***(Raytheon Aircraft Company)***

Theodore "Ted" Wells flew his D4000 to victory in the Portland, Oregon to Cleveland contest. Registered NR6128 (c/n 626), the biplane was raced extensively by Wells during 1928 and 1929 before it was destroyed in a hangar fire at Kansas City. ***(Raytheon Aircraft Company)***

ation with the city's Petroleum Convention. Ira McConnaughey flew R614K and was accompanied on the trip south by Clark in R613K.

The two Travel Airs were the hit of the show and quickly established their dominance over the competition. McConnaughey won the feature event, a 50-mile, closed-course race. James G. Haizlip, who would later share piloting duties with James H. "Jimmy" Doolittle in the *"Shell Mystery Ship,"* also was present at the races. In correspondence with the author in 1983 he recalled learning about the Type R from a pilot who had seen the ship in Wichita, but the Tulsa competition "was the first time I had seen the airplane in flight." Both Haizlip and his wife Mary were well-known pilots on the air racing circuit during the 1930s.

The arrival of cold weather signaled the end of air racing for another year, but there still was a lot of flying

Travel Air employees Dan Doyle (left) and Roy Edwards hold a solid metal model of the Type R that was made to commemorate the victory of R614K in the Thompson Cup race. ***(Dean Edwards)***

ahead for R614K. The airplane had been flown to the New York City area for demonstrations and to allow Curtiss-Wright executives at Garden City a detailed look at the speedster. Following completion of the merger between the Curtiss and Wright companies in October, Walter Beech eventually relocated to New York to assume his new duties as vice president of sales for the Curtiss-Wright Airplane Company. The Wichita factory became known as the Travel Air Division.

In November Doug Davis sought permission from the company to fly the airplane from New York City to his native Georgia to participate in the Atlanta Air Races scheduled for November 9-10, 1929. Curtiss-Wright agreed to the proposal and Doug left New York City at 6:45 A.M. on November 7. After departing New York Davis flew at an altitude of 2,000 feet until reaching Anderson, South Carolina where it began to rain. Undaunted by the weather and anxious to set an intercity record, Doug pressed onward.

The United States Light House air mail radio service kept officials in Atlanta informed of Doug's progress along the route. He was reported leaving Richmond, Virginia at 9:19 A.M., overflying Greensboro, South Carolina at 10:30, and departing Spartanburg, South Carolina at 10:50. When he reached Commerce, Georgia the rain had intensified and navigating by visual contact with the ground became increasingly difficult. As Davis

In January 1930 NR614K was placed into service with the Curtiss-Wright Exhibition Company to perform at airshows and other aviation events throughout the United States. Note the Curtiss-Wright logo on the fuselage below the cockpit. ***(Kenneth D. Wilson)***

later told reporters, "I couldn't have stood the rain or seen a thing if I had stuck my head out" (of the cockpit). "The windshield was so wet I couldn't see through it. Visibility was about 50 feet and I had to fly by compass."

With the weather deteriorating, Doug had decided to turn back to Spartanburg but convinced himself that he could make it to Candler Field near Atlanta. He was gradually forced down to an altitude of 700 feet. Fortunately, Davis avoided obstructions and located the airport where he landed without incident and placed the Type R in his company hangar. He arrived at 11:45 A.M. after flying 4 hours, 30 minutes and covering 785 miles at an average speed of 174.4 mph. Two stops were made for fuel, one at Richmond and the other at Spartanburg, bringing total elapsed time for the trip to five hours. Tired and relieved to be on the ground, Davis credited the Type R for the record flight. "She's a great little ship," he told reporters.

When asked whether he thought Curtiss-Wright would place the monoplane into production following the merger, Doug replied that the company still planned to do so but that a second cockpit would be added and the wing span increased to reduce takeoff and landing speeds. Those plans, however, were dashed in the coming months as the economy collapsed and sales of new airplanes nearly came to a halt.

Davis also believed the Type R had potential as a military fighter and high-speed reconnaissance platform. At times, however, his optimism led to irrational statements. For example, he told the *Journal* that, "if the Mystery Ship had been in the American airports in the World War it could have handled the whole German air force almost by itself, for it could have dropped into a whole squadron of old-fashioned ships and wiped out the entire bunch with little or no danger to itself." And later he opined, "For observation purposes it would be ideal. It could shoot over enemy territory at 20,000 feet to the spot to be observed, make all the pictures required, and get back home almost before an enemy could gain enough altitude to start a battle." Obviously, Davis neglected to consider that if the Americans could produce such an airplane in 1917, the Germans probably could, too.

Despite its many positive attributes, in commercial configuration the Type R was ill-suited for adaptation to military service. The weight of armament and other systems would have reduced overall performance, perhaps to a significant degree. Its influence on future fighter design, however, was undeniable. The Boeing P-26 fighter of 1931 reflected many of the Type R's features with its wire-braced, low-wing configuration and fixed, streamlined landing gear. In addition, after the 1929 Cleveland races biplanes gradually disappeared from the free-for-all competition and by the early 1930s these races were dominated by monoplanes that borrowed heavily from Herb Rawdon's basic design. These included the famous Wedell-Williams racers created by "Jimmy" Wedell and the Granville Brothers' legendary *"Gee Bee"* speedsters.

During the Atlanta airshow Davis flung R614K around the pylons in a crowd-pleasing demonstration of the airplane's brute power. He followed that routine with a dazzling display of aerobatics that included vertical maneuvers as well as aileron rolls and loops. To top off his performance, Doug dove the Type R to an indicated airspeed of about 250 mph., leveling off momentarily in front of 10,000 spectators before hauling back on the stick and climbing straight up to an altitude of 3,000 feet. The *Journal* summed it up well: "All who have seen the ship agree it is a marvel of the age, and many who watched it doing stunts over Candler Field hardly believed their eyes." Such was the mystique of the Type R.

In 1930 the famous racer was flown extensively by Jackson (and occasionally by Clarence Clark) at various aviation events. The airplane was photographed awaiting its turn to make another demonstration flight. A parachute is on the ground. ***(Kenneth D. Wilson)***

Pilot Dale "Red" Jackson poses for the camera before departing for an exhibition flight in NR614K. The Curtiss-Wright logo is visible on the fuselage, and the cowling had been changed to a four-piece unit. ***(Kenneth D. Wilson)***

When the Atlanta event concluded the airplane was returned to Curtiss-Wright. After another minor refurbishing it was transferred to the Curtiss-Wright Exhibition Company on January 27, 1930 for the purpose of public exhibitions and promoting the parent company's commercial airplanes. Since this activity often included air racing, on February 17 the Department of Commerce Aeronautics Branch inspector J.R. Puckett issued another "Restricted" license for R614K that was valid until September 1, 1930.

As part of its transfer from Travel Air the to the exhibition company the airplane's original Wright J-6-9 (serial number 10688) had been removed and replaced by another J-6-9 (serial number 10525). The radial was housed within a new or modified cowling that apparently was installed at the same time as the engine. In addition to the powerplant change, the monoplane was given a complete inspection and thoroughly serviced by the Curtiss-Wright Flying Service in Valley Stream, New York.

Cockpit duties were turned over to Dale "Red" Jackson, an experienced pilot who had gained fame as an endurance flyer in August 1929. He and fellow pilot Forrest O'Brine kept a Curtiss *"Robin"* cabin monoplane orbiting above St. Louis for a remarkable 420 hours, 21 minutes, 30 seconds (17 1/2 days). The two men, who were both employed by the Curtiss-Robertson company, earned $31,000 from supporters and admirers during their aerial marathon. The rash of endurance flights that followed Jackson and O'Brine's stunt captured the public's imagination and prompted the *New York Tribune* newspaper to proclaim that, "Events are

On May 30, 1930 NR614K was photographed after a flight at Logan Airport near Baltimore, Maryland. ***(W.O. Collier Collection)***

Another candid view of NR614K at a regional airshow in 1930. The wheel fairings show evidence of hard use stemming from operations on rough, unimproved grass airfields. Internal braces for the engine cowling also are visible. ***(Antique Airplane Association, Travel Air Club via Robert Taylor)***

In August 1930 the Travel Air was severely damaged in a landing accident near Des Moines, Iowa. The right landing gear was demolished, the right wing was crumpled, and the vertical stabilizer was smashed. ***(John Sunyak)***

coming so thick and fast that it is difficult to keep track of them. The air today is like a circus, so full of wings it bewilders the mind." [6]

Clarence Clark occasionally flew the ship, but it was Jackson who performed the majority of the flying. During the next eight months the Travel Air visited airshows and exhibitions across the country performing aerobatics and high-speed dashes that thrilled spectators and helped them forget, if only for a few minutes, the worsening economic woes that were befalling the nation. These exhibitions also carried the flag for the struggling Curtiss-Wright Corporation as well as the Gulf Oil Company, which sponsored the airplane to advertise its *"Gulfpride"* aviation gasoline and oil products.

The monoplane, which had been issued a special "NR" license by mid-1930, was kept busy attending many events and shows in 1930 and eventually the arduous schedule took its toll on the machine. For example, in March minor damage to the tail surfaces were repaired, and in May the ship underwent a factory overhaul that included recovering the wings, replacing all bushings, installing new wheels and axles, and fuel tanks. There were no major repairs made in June or July.

In August Jackson was replaced as pilot for R614K by J.L. McGrady. According to racing pilot James Haizlip, McGrady was an "old Issoudun pal who was making an effort to get back into gainful aviation." He had flown French Nieuport fighters in World War One and was known to be a "smooth pilot, but how much he had kept his hand in during the years between 1918 and 1930" was unknown, Haizlip wrote to the author in 1983. McGrady may not have been familiar with high-performance airplanes such as the Type R, for on August 28, 1930 he made a disastrous landing with the ship at the Iowa State Fair near Des Moines.

One source states that during the landing the right gear struck a hole in the ground, causing the gear to collapse. The propeller and cowling dug into the dirt and the racer went over on its back. Another source, however, recalled that McGrady landed the airplane at high speed and struck the ground hard, bouncing several times before he totally lost control and the aircraft went inverted, sliding tail-first to a stop. Regardless of which version is true, the airplane was severely crippled. The aft fuselage and the empennage were crushed, the right wing was crumpled, and the propeller, engine, and NACA cowl were damaged. An embarrassed McGrady escaped with only minor injuries, but R614K was seriously wounded.

After the accident at Des Moines NR614K was stripped of useable parts and components and placed in storage to await its fate. Two fuel tanks are visible in the photograph; one is under the trailing edge of the left stub wing, and the other is upright and leaning against the stub wing. A damaged wheel fairing is in the foreground along with various sheet metal pieces. ***(Kenneth D. Wilson)***

Curtiss-Wright officials were undecided whether to repair the airplane and it was transported to a company hangar at Lambert Airport in St. Louis and placed in temporary storage. By late in 1930 sales of new and used airplanes were deplorable and getting worse by the week, and Curtiss-Wright management had no intention of spending its financial resources to repair the abandoned racer. But a young pilot named Walter Hunter would soon rescue the famous Travel Air from oblivion and give it a new lease on life.

Hunter was one of four brothers that established a world record for endurance flying in 1930. All of the Hunter boys had been bit by the flying bug, and to make a living they barnstormed the mid-west in their dilapidated Standard J-1 biplane and billed their troupe as the "Hunter Brothers Flying Circus." After noting the financial success of Jackson and O'Brine's record the previous year, the four farm boys from Sparta, Illinois had pooled their meager cash reserves and bought a weary Stinson Detroiter cabin monoplane and overhauled its Wright J-5 radial engine in preparation for the own airborne marathon.

On June 11 John and Kenneth Hunter took off from Chicago's Sky Harbor Airport to begin what would prove to be a grueling ordeal in the air. They would be resupplied and the airplane refueled by brothers Walter and Albert flying the old Standard that had been converted into an aerial tanker. The Stinson, christened *"City of Chicago"* by the boys, droned on and on above the airport as time dragged by.

After nearly 20 days aloft their stunt began to pay off as they passed Jackson and O'Brine's mark, and companies began to offer cash and other incentives for each additional day they remained in the air. The Stinson had been airborne for 24 days when a ruptured oil line ended the brothers' quest for glory. The Stinson *"City of Chicago"* landed safely after flying for 553 hours, 41 minutes. 30 seconds. [7]

Walter Hunter sits in the cockpit of the rebuilt NR614K at Lambert Airport in St. Louis, Missouri before departing to compete in the 1931 Bendix Trophy race. Mechanic Herbert Bud, who supervised the extensive rebuilding project for Hunter, is standing with his back to the camera. ***(Courtesy of the Hunter Family via Jack and Pauline Winthrop)***

After the endurance flight Walter Hunter eventually became a pilot with Universal Airlines. Although he had a good flying job and a steady paycheck, he was attracted to the large purses being offered for the upcoming 1931 Thompson Trophy and Bendix Trophy races (the Thompson competition was a closed-course event, and the Bendix was a free-for-all, cross-country race). With thousands of dollars up for grabs, Hunter began laying plans to compete in both contests and initiated a search for an airplane that was suitable for the task. Building a new racer was out of the question. But he knew that Curtiss-Wright still had the wrecked Travel Air Type R and Hunter was confident that it could be rebuilt into a competitive machine.

According to Federal Aviation Administration records for R614K, on December 22, 1930 Curtiss-Wright sold the wreck to Hunter. Another source, however, claims Walter bought the airplane in June 1931. But that date would allow only three months to rebuild and test the ship before the start of the Bendix race, scheduled for the first week in September. Considering these factors, it appears that the December 1930 date is correct.

Although Walter had an airplane, much work lay ahead before it could be returned to an airworthy condition and transformed into a serious contender for the Thompson and Bendix crowns. What Walter needed was an expert mechanic to resurrect R614K from its ashes. Fortunately, he had the right man in mind—Herbert Bud. Bud was an experienced technician who worked for the Parks Air College in St. Louis. The school was well-known and by 1930 had earned an excellent reputation for training pilots and mechanics.

Hunter discussed his project with Parks officials and they agreed to allow the ship to be rebuilt in the school's shops using student labor under Bud's direct supervision. As a result, the airplane was transferred to Curtiss-Steinberg Airport. The airport had begun operating in 1928 and was located less than one mile northeast of the Parks school. By 1930 the name had been changed to Curtiss-Steinberg Airport to reflect the Curtiss-Wright Corporation's controlling interest in the facility. In the 1980s it was renamed "St. Louis Downtown Parks Airport".

Soon after R614K entered the shops the repair and modification process commenced. Hunter had a few modifications planned for the would-be racer and these were engineered by William Thompson, who also worked for the college. The chief alterations included a three-piece, quasi-canopy enclosure and the paint scheme was changed to black overall with orange trim. To power the ship for its initial test flights Hunter installed the Wright J-5 engine removed from the *"City of Chicago."*

According to notes prepared by Thompson dated August 1931 the fuselage and both wings had been repaired and a new fuselage fairing installed. The entire airplane was recovered with fabric, and aileron bushings were replaced. Having risen from its grave like the mythical Phoenix, on August 6, 1930 Travel Air R614K took to the air nearly one year after the accident near Des Moines. Walter Hunter flew a brief series of tests to check the airplane and its systems before departing for Wichita, Kansas to have a new NACA cowling installed that had been built by Herb Rawdon and Walter Burnham. In mid-August the assembly was installed by the Christopher Airplane Service whose mechanics also installed a 20-gallon fuel tank in the fuselage, re-rigged the landing gear and the vertical and horizontal stabilizers, rebuilt the tail skid, installed a new windshield, streamlined the flying and landing wires at the wing fittings, and installed a new brace for the instrument panel.

Hunter flew back to St. Louis before departing for northern New Jersey where Wright Aeronautical Corporation removed the old J-5 radial and installed a supercharged SR-975E engine (serial number 12569) that had been modified to produce more than 500 hp. The engine already had been run-in on the test stand for 12 hours. With race day less than a week away Walter took off and headed west, arriving in Los Angeles on August 30 to begin final preparations for flying in the Bendix contest. On August 31 the engine was given a detailed inspection by mechanic Fred Matthews, and during the next few days Hunter made a series of short flights to check the engine and all systems.

The 1931 Bendix event offered a first place purse of $7,750— a lot of money in the Great Depression. Eight pilots would be competing for the prize, including Jimmy Doolittle in a Laird *"Super Solution"* biplane. In the early morning darkness of September 4 the airplanes were ready to launch from the Union Air Terminal in Burbank, California. In addition to Doolittle's fast Laird and Hunter's born-again Type R, there were three Lockheed Altairs and two Orions (retractable landing gear) and one Vega. The big Lockheeds carried additional fuel tanks and were capable of flying the race to Cleveland non-stop, but the two custom-built machines would have to land for fuel along the way.

Poised at Burbank, California before the start of the 1931 Bendix contest, NR614K was photographed undergoing a calibration of its magnetic compass. Note race number "16." A new Wright J-6-9 rated at more than 500 hp. was installed in the ship along with a new cowling fabricated by Herb Rawdon and Walter Burnham. **(John W. Underwood)**

DOUGLAS H. DAVIS

An Air Racing Legend

Born and reared in Georgia, Davis learned to fly during World War One. In 1918 he joined the U.S. Army Air Service and was posted to Maxwell Field in Montgomery, Alabama where he soloed on November 17 in a Curtiss JN-4 biplane. He was transferred to Carlstrom Field in Arcadia, Florida where he served as a flying instructor and check pilot until his discharge in May 1919.

Doug returned to his hometown of Griffin, Georgia and soon purchased a war-surplus JN-4 that he used to haul light cargo and give sightseeing flights. Davis also barnstormed throughout the region before joining the famous Cody Flying Circus in Florida in 1923. Late that year he had relocated to Detroit, Michigan and began giving flying lessons in his war-weary JN-4 *"Jenny."* He bought a second airplane in 1924 to help fulfill contract obligations to promote the *"Baby Ruth"* candy bar made by the Curtiss company. In 1925 he purchased the first of two Waco 9 biplanes that he flew to promote the candy in 38 states.

Davis married his high school sweetheart, Glenna Mae D'Hollosy, in December 1925 and the couple took off in the Waco on a 11,000-mile honeymoon. The next year he began commercial operations as the Doug Davis Flying Service in Atlanta, and gained valuable experience as a neophyte race pilot by competing in the 1926 National Air Races near Philadelphia, Pennsylvania. Although Doug had been a successful Waco dealer, in 1928 he became a distributor for the Travel Air Company.

Being familiar with Davis's flying abilities and air racing experience, Walter Beech tapped his friend to fly the Type R in the 1929 National Air Races at Cleveland. Doug's victory thrust him into the spotlight as one of America's best pilots. Between 1929 and 1934 Davis continued operating his flying business and competed in a number of major air races. During the 1932 National Air Races he won the Sohio Mystery Derby flying his Travel Air biplane—the little B11D built in 1929 as the third in a trio of racers that included the two Type R ships. He also took first place with the ship in Event No. 12 for 180-mph. airplanes. Doug had bought the airplane and modified it over the years to make it a legitimate competitor. It was powered by a Wright J-6-7 of 240 hp.

In July 1934 he was asked by Harry Williams, a partner in the Wedell-Williams Air Service in Patterson, Louisiana, to fly the famous Wedell No. 44 machine in the upcoming National Air Races. Davis would replace pilot "Jimmy" Wedell who had been killed in June while giving flight instruction to a student pilot.

On August 27 Doug took off in No. 44 and flew to Burbank, California to fly in the Bendix cross-country race to Cleveland, Ohio. Although the long flight went well, Doug realized that the elevator trim system was not working properly. Mechanics checked the jackscrew that moved the horizontal stabilizer up and down, and thought they had eliminated the problem. On August 31 No. 44

A young Doug Davis poses with a pair of Waco 9 biplanes he flew to advertise the popular "Baby Ruth" candy bar for the Curtiss company. To publicize the product Davis would drop small samples that were tied to tiny parachutes. The crowds below scrambled to grab the treats. ***(Kenneth D. Wilson)***

Davis eventually purchased the B11D racer built by Travel Air for the 1929 National Air Races. He modified NC612K over the years and it became a frequent participant in local and regional airshows in the southern United States. **(Kenneth D. Wilson)**

Davis was killed flying the Wedell-Williams racer No. 44 during the 1934 Thompson Trophy race. He had won the 1934 Bendix contest in the same airplane. **(Kenneth D. Wilson)**

departed Burbank and was first to arrive at Cleveland. Average speed for the race was 216.2 mph. During the flight, however, Davis again had difficulty with the stabilizer trim system and again repairs were made in an effort to resolve the problem.

At Cleveland Davis won the race for engines of 1,000 cubic-inch displacement as well as the "Shell Speed Dash" competition at an average speed of 306.2 mph. Next, he prepared for the feature event—the Thompson Trophy contest. Doug led the race and was widening the distance between his black and red racer and another Wedell-Williams flown by Roscoe Turner. On the eighth lap as Davis streaked down the backstretch and circled the second pylon, spectators were horrified to see the airplane suddenly enter a spin and dive nose-first into the ground behind a copse of trees. Davis was killed instantly. The cause of the accident was never determined, but gave rise to suspicions that the stabilizer trim system had failed and contributed to the crash.

Not long after the races Mrs. Davis received a package from some of Doug's friends who were at Cleveland when he died. She opened the box to find her husband's soft, Panama hat that he had adorned with the autographs of 38 aviation luminaries including Alford Williams, Clyde Pangborn, Ben Howard, and James and Mary Haizlip. Davis had worn the hat during the races. In 1970 the street near Atlanta's sprawling Hartsfield International Airport was named "Doug Davis Drive" in honor of the city's most famous aviator who had started a flying service at a small, grassy field nearly a half century earlier. In 1972 Davis was inducted into the OX-5 Club of America Hall of Fame.

At 1:20 A.M. Larry Therkelsen, the official starter for the National Aeronautic Association that sanctioned the race, dropped his flag and one of the big Altairs roared down the runway and into the night, heading east for Ohio. About 15 minutes later the exhaust noise from R614K's fire-breathing Wright ripped the air asunder as the airplane sped down the strip and was off on its bid for the Bendix Trophy. During the next 20 minutes all of the racers took off at intervals.

Hunter climbed his powerful steed to an altitude of about 18,000 feet and settled down for the flight to his first fuel stop at Amarillo, Texas. Walter was suffering from a head cold and his ears hurt, but he forged ahead until a problem with fuel flow to the engine forced him to land prematurely at Winslow, Arizona. The trouble was soon traced to a fuel leak in one of the lines. After repairs Walter refueled and departed for Lambert Field in St. Louis. Unfortunately, misfortune struck him again, this time over Kansas. The J-6-9's heartbeat became erratic for the second time, and Hunter was forced to make a precautionary landing at the U.S. Army's Ft. Riley facility where he was able to refuel and take off for St. Louis. He knew the unscheduled stops probably had erased any hope of winning the Bendix contest, but if he could stay airborne and refuel quickly at Lambert Field, there was at least a slim chance of placing second and collecting $3,750.

True to their promise, the ground crew at Lambert Field was ready and made short work of refueling the Travel Air. Walter was aloft again and heading for the finish line at Cleveland. If he had made it to the airport Hunter would have clinched second place, but near Terra Haute, Indiana the nine-cylinder Wright suddenly began to cough and sputter. Walter worked feverishly to restore full power, but it was not to be. Hunter skillfully set the sick racer down in a field, and as he climbed out of the cockpit to assess the situation smoke and flames were licking around the edges of the cowling. Fortunately, a local fire department soon arrived on the scene and extinguished the minor blaze without seriously damaging the ship. The magnetos, however, were burned badly. After being notified of Hunter's predicament, the Wright company dispatched a technician and a pair of magnetos to Terra Haute to make repairs.

Early the next morning, September 6, Walter took off from the field and finally arrived over the municipal airport at Cleveland about one hour later. He had lost the Bendix (Doolittle won) but was optimistic that R614K still could win the Thompson race. James Haizlip, who had made a tentative agreement with Hunter to fly the airplane in the Thompson contest, was attending the races and witnessed what happened next. As Haizlip recalled, "I was there with Mary (his wife) who was trying out a Laird "Sportster" belonging to a Memphis friend who was lending it to us for some of the closed-course events for airplanes with Approved Type Certificates. When Hunter approached from the east I recognized the airplane. He continued to make a counter-clockwise circle of the race course, a sensible maneuver, then approached from the north on final for a landing on the north-south runway."

In a letter to the author Haizlip wrote that for the 1931 races much of the airport was surrounded by a high canvas wall emblazoned at intervals with the word *"STANAVO."* It was erected to prevent non-paying spectators from seeing the races, especially the events flown at low-altitude, for free. Haizlip continued: "As the Type R came over the trees at about 150 feet above the ground and a furlong short of the canvas wall, suddenly a parachute bloomed and the airplane disappeared followed by a black pall of smoke that erupted and obscured the parachute settling limply out of sight. At the time I was at least 400 yards away. I could only visualize my friend lying crushed to death behind that brown canvas curtain. The sirens of the field ambulance and fire equipment reminded me that there was nothing I could do, so with a heavy heart, somewhat conditioned to accidents by the high attrition rate during World War One, I did my mourning silently."

"It was only later in the day when a report came through that Walter Hunter was alive and being treated at the Berea, Ohio hospital that my spirits rose. That evening I drove down to see him. To my delight, there was Walter sitting up in bed with both hands bandaged, his sandy hair and ruddy complexion heightened a bit as if by a severe sunburn." Haizlip sat down to talk with Hunter, who was eager to share his story of the day's events.

In further correspondence with the author, Haizlip

Walter Hunter poses with NR614K. Engine problems caused by loose fuel line fittings would end his bid to win the 1931 Bendix race. Hunter later was forced to bail out of the burning airplane on final approach to the airport at Cleveland, Ohio. ***(John W. Underwood)***

wrote that he carefully recorded what Walter had said that night and it makes for an interesting commentary: "When I left Terra Haute this morning I thought everything was go. The broken gas line had been replaced and the engine was running OK. My head cold was still bothering me. I was stuffed up and couldn't smell or breathe through my nose, but otherwise I figured my troubles were over. The other times that I had been forced down was on account of the gas lines parting at the flared fittings. I was convinced that we had that licked."

In addition to the design of racing airplanes, the Travel Air Type R also influenced future fighters developed for the United States Army Air Corps. In 1931 Boeing introduced an open cockpit, low-wing, wire-braced monoplane with a fixed landing gear that was powered by a Pratt & Whitney radial engine. It was designated P-26A by the Army. The overall resemblance to the Type R is obvious. ***(W.O. Collier Collection)***

A replica of R614K was built late in the 1970s by Jim Younkin. It was powered by a Lycoming radial engine and was a popular attraction at regional and national airshows. ***(Antique Airplane Association via Robert Taylor.)***

"As I arrived at Cleveland I was paying attention principally to the traffic around the airport and as I recall, hadn't looked down into the fuselage until turning from base leg to final when I felt heat on my lower legs. I looked down, and in the bottom of the plywood fuselage there must have been several inches of gasoline sloshing around and I hadn't smelled it. As I shut off power for the final glide, the flare from the short engine [exhaust] stacks must have ignited the liquid and flames were up around my knees. I was still up about 200 feet but too low to jump. For an instant I was ready to give up. Then I thought of our widowed mother down home (at Cairo, Illinois) and decided to give it at least a try."

"I was wearing a parachute that the rigger in St. Louis, who claimed he was the best in the business, had packed just two weeks before. As I rolled over the left side, pulling the cord as I cleared the cockpit, you wouldn't believe it! The forward motion of about 70 mph. filled the chute instantly as I swung almost horizontally forward of the canopy. Apparently, at the precise instant that the first backswing would have started, my feet touched the ground as the airplane and engine buried itself in the turf a couple of hundred feet ahead and burst into flame. You won't believe me, but the landing for me was no more than stepping off a chair a couple of feet to the floor."

The crash brought a ignominious end to the career of Travel Air R614K. As for Hunter, he spent the next month in the hospital recovering from burns before returning home to Sparta, Illinois. By October he was back in the cockpit flying for Universal Airlines. In 1966 Hunter retired from Universal's successor, American Airlines, as a senior captain flying the Boeing 707. James Haizlip remembered that the last time he saw Walter was 35 years after the crash at Cleveland when he and a friend were guests at the Haizlip's home in Pacific Palisades, California. Hunter died October 17, 1983 in St. Louis, Missouri. [8]

In the short span of two years Herb Rawdon and Walter Burnham's Type R had become an icon in American air racing. The little monoplane was a testimony to their design genius and guaranteed both men a well-deserved place in aviation history. More importantly, its success signaled the birth of a new and exciting era in the design and performance of both commercial and military airplanes, especially those built expressly for air racing. Perhaps Doug Davis best described the enduring mystique of the Type R when he said that day in November 1929, "She's a great little ship." [9]

CHAPTER THREE NOTES

1. It is uncertain whether canvas covers were thrown over the two racers after they were placed in the hangar. It is possible, since Walter Beech ordered strict security to protect the airplanes from the prying eyes of competitors and the public. Travel Air employee Theodore "Ted" Cochran kept a watchful eye on the two ships whenever they were in the hangar.

2. Bowers, Peter M. *"Curtiss Aircraft - 1907-1947"*; Putnam & Company, Ltd., London, 1979.

3. In 1929 the National Aeronautic Association (NAA) granted Charles Edwin Thompson's request to offer a permanent trophy each year for the free-for-all event at the National Air Races. The striking Thompson Trophy was a figure of mythological Icarus and stood 40 inches in height. It was made of gold and silver mounted on a marble base. A model of each year's winning airplane would be placed at the top of the trophy.

   Similarly, in 1931 the NAA authorized Vincent Bendix, president of the Bendix Aviation Corporation, to sponsor a trophy to be awarded to the winner of a free-for-all transcontinental speed race between Los Angeles and Cleveland. The chief purpose of the contest was to, "encourage experimental work toward the development of higher cross-country speeds and new transcontinental speed records," according to the Bendix company. James H. Doolittle won the first Bendix competition in 1931 in a Pratt & Whitney-powered Laird biplane. James Haizlip won the race in 1932 flying a Wedell-Williams racer. In 1936 aviatrix Louise Thaden accompanied by Blanche Noyes became the first female pilot to win the prestigious event. Her airplane was a Beechcraft Model C17R.

4. According to Don Berliner, an expert on the history of the Cleveland air races, the official speeds and elapsed times for Event No. 26 have been questioned by individuals who claim the figures are inaccurate. Their calculations and conclusions regarding speeds and times of the race contradict official observations.

5. By the late summer of 1929 the Travel Air factory was producing at least 25 airplanes per week and was struggling to attain a rate of nearly 50 per week. Although the ratio of open cockpit biplanes to cabin monoplanes being built was about 60% to 40%, respectively, the popularity of cabin airplanes was increasing.

6. Dwiggins, Don; *"Famous Flyers And The Ships They Flew,"* Grosset & Dunlap, Inc., New York, 1969.

7. In July 1935 brothers Fred and Al Keys flew their Curtiss monoplane dubbed *"Ole Miss"* to a record endurance time of 653 hours, 34 minutes, or more than 27 days aloft. By the mid-1930s public enthusiasm for such stunts was waning and the flights ended.

8. The remains of R614K were stored in a hangar at Cleveland until after the races when some parts and components were salvaged. Eventually, the tail section found its way to the Staggerwing Museum Foundation in Tullahoma, Tennessee where it was placed on display along with other minor components from the airplane.

9. The Staggerwing Museum Foundation in Tullahoma, Tennessee commissioned construction of a replica Type R bearing the registration R614K. The airplane was a faithful representation of the original ship and was undergoing flight testing in October 1995 when it crashed while landing at the Youngstown Elser Metropolitan Airport in North Lima, Ohio. The pilot, Robert C. Van Ausdell, died from injuries received during the accident.

CHAPTER THREE

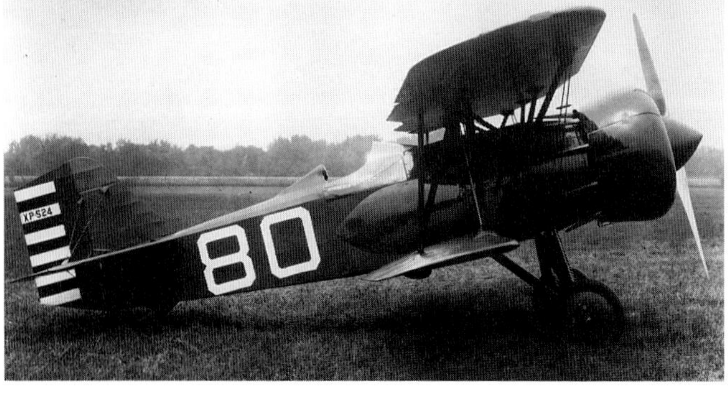

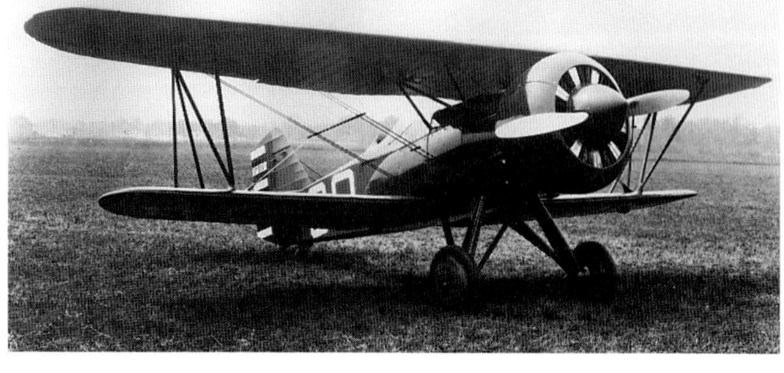

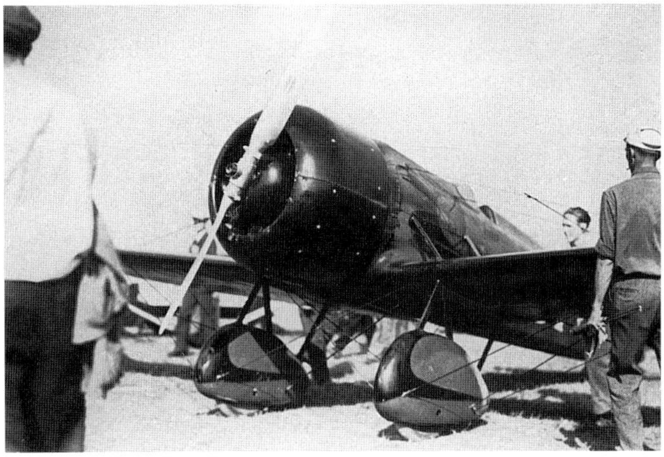

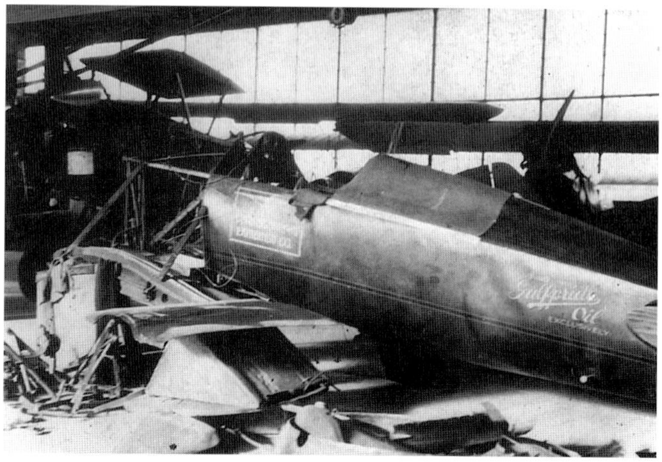

Curtiss-Wright officials were undecided whether to repair the airplane and it was transported to a company hangar at Lambert Airport in St. Louis and placed in temporary storage. By late in 1930 sales of new and used airplanes were deplorable and getting worse by the week, and Curtiss-Wright management had no intention of spending its financial resources to repair the abandoned racer. But a young pilot named Walter Hunter would soon rescue the famous Travel Air from oblivion and give it a new lease on life.

Hunter was one of four brothers that established a world record for endurance flying in 1930. All of the Hunter boys had been bit by the flying bug, and to make a living they barnstormed the mid-west in their dilapidated Standard J-1 biplane and billed their troupe as the "Hunter Brothers Flying Circus." After noting the financial success of Jackson and O'Brine's record the previous year, the four farm boys from Sparta, Illinois had pooled their meager cash reserves and bought a weary Stinson Detroiter cabin monoplane and overhauled its Wright J-5 radial engine in preparation for the own airborne marathon.

On June 11 John and Kenneth Hunter took off from Chicago's Sky Harbor Airport to begin what would prove to be a grueling ordeal in the air. They would be resupplied and the airplane refueled by brothers Walter and Albert flying the old Standard that had been converted into an aerial tanker. The Stinson, christened *"City of Chicago"* by the boys, droned on and on above the airport as time dragged by.

After nearly 20 days aloft their stunt began to pay off as they passed Jackson and O'Brine's mark, and companies began to offer cash and other incentives for each additional day they remained in the air. The Stinson had been airborne for 24 days when a ruptured oil line ended the brothers' quest for glory. The Stinson *"City of Chicago"* landed safely after flying for 553 hours, 41 minutes. 30 seconds. [7]

an exciting week of racing and the fans were not disappointed.

But nothing attracted their interest more than Event No. 32—the Thompson Trophy Race held on Labor Day. With $5,000 at stake the starting lineup was impressive. "Speed" Holman was in the Laird *Solution,* his 220-pound, 6 foot, five-inch frame shoehorned into the tiny cockpit; Frank Hawks and his Travel Air *Texaco 13* were ready, along with Marine Corps Captain Arthur Page in the mighty XF6C-6, Errett Williams would be flying the Wedell-Williams "We Winc" monoplane, and Paul Adams was perched in his Travel Air D4000 biplane. Ben O. Howard and his pugnacious little "Pete" racer were there, and last but least, James G. Haizlip in the powerful *Shell Mystery Ship.* To give Jim an additional edge in the race, Shell engineers had concocted a special brew of 90-octane gasoline mixed with toluene, an anti-knock additive intended to suppress detonation at high power settings.

The gladiators of the air were ready to tackle the 20 laps over a five-mile, triangular course. After ensuring that all was in order official time keepers launched the airplanes at 10-second intervals. Arthur Page was first to take off in the XF6C-6, followed quickly by the other contestants. He easily took the lead, and it soon became apparent to Haizlip and his fellow pilots that the Navy speedster would win the race unless one of them could overtake and pass the fast Curtiss machine. As for the rest of the contestants, Haizlip's adversary Frank Hawks was forced out of the race on the third lap with fuel flow problems, and Williams retired on the eighth lap.

With Page far ahead of the pack, the real battle was between Holman and Haizlip for second place. "Speed" had taken off 20 seconds behind Shell's ship but as the race neared its end Jim had not yet spotted either Holman or Page on the course. As Haizlip later explained, "...coming down the home stretch with the packed grandstands on my right and the bunting striped pylon on my left, I was concentrating on my turn pattern that I'd worked out and practiced in the previous day's races and convinced myself that it was about the most efficient for this race course that I could work out. Fatigued as I was, this approach toward the crowd never failed to pump a shot of adrenaline into my bloodstream.

Flying at 90 mph., Doolittle and his racer were photographed from a Curtiss Robin chase aircraft. Minutes later Doolittle made a high-speed pass across the airfield. Flutter developed and ripped off the part of the ailerons and wing. Thinking fast, Jimmy pulled up and to the right to avoid the crowds below before rolling NR482N inverted and jumping from the cockpit. He landed safely. ***(Smithsonian Institution negative number 83-3564)***

My right foot was getting numb from general dehydration and the constant pressure of holding top rudder coming out of turns. Just as I started to lean the airplane over gently to show it the home pylon, I sensed a shadow above and to my right. There, big as life, above and outside my turn radius as the rules required, was Speed Holman's Laird, and just passing us both farther to the right and somewhat higher was Captain Page's Curtiss. As Captain Page swung southeastward in a wide turn he pulled up. I watched Page's airplane shadow come up too fast under the gliding, powerless airplane. The instant it hit I could see it stop as it plowed into the hard turf in a space just a little more than the length of the fuselage. No bounce, just deep plowing." Page was severely injured and died the next day. [4]

Stunned by what he had seen, Haizlip fought off feelings of remorse and concentrated on flying the course.

Only a pile of wood, fabric and steel remained after NR482N crashed at high speed. The engine had buried itself about five feet in the ground. The aileron control tube that was redesigned by Doolittle caused the flutter that led to the crash. He blamed no one but himself for the accident. ***(Smithsonian Institution negative number 83-3561)***

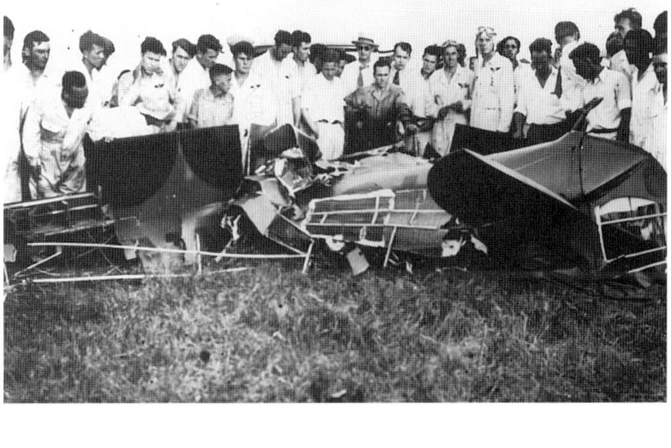

After the demise of his modified Travel Air speedster, Doolittle flew the "Shell Lightning" (NC12222), a Lockheed "Orion" featuring retractable landing gear. ***(W.O. Collier Collection)***

What had been a fight for second place only moments before had suddenly become a dogfight for the checkered flag between Holman and Haizlip. Unfortunately, the fuel mixture prepared for the Shell racer was causing the Wright J-6-9 to loose 50-100 RPM and the engine was smoking heavily. From then on the race was fought on the pylons," Jim wrote in a letter to the author. Exhaust from the engine's short stacks had covered part of the windshield, and the forward fuselage was black with stains.

In the remaining laps Haizlip continued to stay inside Holman during the turns, which gave him a slight advantage in overall speed. But the race was won on elapsed time, not speed. If Jim had been first to cross the finish line he still would have lost the battle. In the end it was Holman who won at an average speed of 201.91 mph. Haizlip was second with a speed of 199.80 mph. and earned $3,000. Ben Howard won $2,000 for third place. Although he was disappointed to have lost to Holman, Haizlip could take some consolation in the knowledge that he had made a valiant effort to win but was handicapped by the improper gasoline mixture. If the engine had been producing full power the outcome may have been different.

After the races concluded NR482N was flown back to the Travel Air factory at Wichita where the 400 hp.

Doolittle's Laird LC-DW-500 "Super Solution." He flew the biplane to victory in the 1931 Bendix Trophy race and also set a transcontinental record. ***(W.O. Collier Collection)***

Wright was removed and replaced by the 300-hp. J-6-9 engine. The colorful *Shell Mystery Ship* was ready to resume its publicity schedule when on October 1, 1930 Doolittle collided with an Army Air Corps trainer in the air above the Kansas City, Missouri airport. Jimmy was in a vertical climb when his Travel Air struck the tail section of the Army aircraft. Although the Army pilot managed to parachute to safety and Jimmy landed without further incident, Doolittle was later held responsible for the collision and fined $100.00. In a letter dated October 22, 1930 Gilbert G. Budwig, director of Air Regulation for the Aeronautics Branch, acknowledged receipt of a personal check sent by Jimmy along with an explanation of his "violation of the Air Traffic Rules at Kansas City, Missouri." The yellow and red racer had incurred damage to the landing gear and wings but was quickly repaired and was soon flying again. [5]

Unfortunately, its rejuvenation was short-lived. on November 16, 1930 Haizlip was departing Lambert Field in the monoplane when the engine suddenly failed, possibly because of fuel starvation. He recalled that "In my desperate effort to save the airplane in spite of the obstructions ahead and on both sides, and a headwind of 30 knots, I managed to land directly below between hangars without turning back only to pitch headlong into a wide drainage ditch just inside the field boundary, and nose into the ditch bank at 35 mph. to mash my face and break the poor aircraft into three neat parts. I recovered the next day but the aircraft became scrap." [6]

The Shell Oil Company decided not to rebuild the wreck and the registration NR482N was cancelled by the Aeronautics Branch on November 24, 1930. The reason for the cancellation was listed simply as "washout." The once-proud *Shell Mystery Ship* was relegated to storage to await final disposition. Jimmy Doolittle, however, saw a potential record setting speedster amidst the broken fuselage and wings. After discussing the matter with his wife and negotiating with Shell, he dug deep into the family's savings and bought the crippled Type R in April 1931. Doolittle had set his sights on establishing a speed record for land-based airplanes, and he planned to put his aeronautical engineering knowledge to work by making modifications to the ship that would transform it into a racer capable of more than 300 mph. At that time the record in America stood at 266 mph. and the world record was 277 mph.

He made arrangements with Parks Air College in St. Louis, Missouri to completely rebuild the aircraft to his specifications and the airframe was transported to facilities at the Curtiss-Steinberg Airport in nearby Cahokia, Illinois where the work was performed. During the next three months a new fuselage was built incorporating structural changes that eliminated the need for external cabane struts between the fuselage and the stub wing. Jimmy also engineered a system that retracted the wheels into large fairings after takeoff. The empennage, which was relatively intact after Haizlip's crash, was grafted onto the fuselage and a pair of short-span wings were installed on the new machine.

In a letter to the author dated January 10, 1984

Doolittle wrote that "The original big wings were destroyed in Haizlip's accident. I had the plane rebuilt around the small wings, which were almost new." He was referring either to the set of racing wings originally built for R614K, which Walter Burnham claimed Doolittle had obtained from Travel Air, or possibly to a different set built specifically for the original *Shell Mystery Ship*. In the letter Doolittle did not indicate which wings were installed on the rebuilt airplane. In addition, Department of Commerce records noted only that Doolittle "had an extra set of wings" that were installed on the ship.

To power his racer Jimmy installed a modified Pratt & Whitney *"Wasp Jr."* rated at 560 hp. at 2,450 RPM. and enclosed the engine within a new cowling. A Hamilton Standard aluminum alloy propeller (hub serial number 7396) with a diameter of 94 inches was fitted to the engine and the blades were set to an angle of 24 degrees. The gross weight of the airplane was 2,500 pounds. In the third week of June 1931 Aeronautics Branch inspector Guy Faulkner approved the modifications and Jimmy began preparations for the airplane's first flight. In addition, Doolittle had applied for and received the registration NR482N from the Aeronautics Branch. Although inspector Faulkner had approved the new monoplane he specifically requested that Doolittle test-fly the powerful ship before he would issue the license card. It proved to be a wise decision.

On June 23, 1931 Jimmy and NR482N were ready to fly and the usual first-flight crowd gathered at the airport to witness the event. Among them was 10-year old James Doolittle, Jr. After donning his parachute Doolittle climbed into the cramped cockpit. The powerful Pratt & Whitney radial belched smoke and shuddered as all nine-cylinders came to life and the airplane taxied for takeoff. Minutes later the roar of the engine alerted observers that the ship was rolling down the runway and accelerating rapidly. It seemed to leap off the ground and the initial climb was described by observers as "spectacular." During the first few minutes of flight Jimmy checked handling characteristics and performed a few aerobatic maneuvers. All went well.

Next, he applied full throttle and determined that the maximum indicated airspeed was 250 mph. In accordance with a pre-arranged plan, Doolittle then slowed down to 90 mph. and for the next few minutes flew formation with a Curtiss *Robin* cabin monoplane while a photographer snapped aerial views of the reborn NR482N. After the formation flying was completed Doolittle climbed for altitude and began a gentle power dive. As the ship leveled off about 100 feet above the runway at a speed estimated to have been 250 mph., one spectator later recalled that it looked as though the airplane had suddenly "struck a bunch of birds."

But instead of feathers it was pieces of wood and fabric that were flying off the right wing as the aileron departed followed moments later by parts of the left aileron. Jimmy later recalled that "I suddenly felt the right wing become heavy. I couldn't tell what had happened but I knew I would have to nose her up if I was going to get out. I pulled the ship up as steep as I could and turned her over on her back. Then I let loose. I was so close to the ground it seemed I didn't have time to do more than pull the ripcord before I landed."

In the early 1930s the "Gee Bee" racers built by the Granville brothers took the Travel Air Type R's design philosophy to its ultimate expression. Doolittle won the 1932 Thompson Trophy Race in the R-1 (NR2100), photographed at Cleveland, Ohio in September 1932. ***(W.O. Collier Collection)***

After pulling up and to the right to avoid the spectators, Jimmy had abandoned the stricken racer at an altitude of about 300 feet. It was one of the lowest jumps on record. He landed hard and suffered a rope burn across the back of his neck, but was uninjured. When Doolittle bailed out observers on the ground said the airplane "seemed to stand for a brief second in mid-air, then fell straight and heavily to the ground." The engine buried itself five feet in the sod and the airframe was almost totally obliterated by the force of impact. Only 12 seconds had elapsed between the onset of flutter, aileron failure, and Jimmy's exit from the cockpit. The parachute that saved his life had been repacked one week earlier by sergeant James Tate of the 35th Division Air Service, Missouri National Guard. He also had packed a parachute that was used in an emergency escape by Charles A. Lindbergh.

Doolittle knew what had caused the crash and blamed

James Haizlip won the 1932 Bendix Trophy Race in the Wedell-Williams No. 92 (NR536V) and beat Jimmy Doolittle's 1931 transcontinental record with a time of 10 hours, 19 minutes. ***(W.O. Collier Collection)***

no one but himself: "The wing failure was all my own fault as, due to a change in the wing fairings, I designed in bent torque tubes which permitted some play and encouraged wing flutter," he wrote to the author in January 1984. It is interesting to note that Jimmy told reporters he had taken the wings from "an old plane," and *Aero Digest* magazine reported that Doolittle had "used a pair of racing wings from a previous speed plane and apparently they were too old or weak to stand the strain he imposed on them."

The Aeronautics Branch cancelled registration NR482N on July 18, 1931. Following the crash Doolittle found the ripcord from his parachute and in the tradition of those who became members of the *"Caterpillar Club,"* he kept it as a souvenir. It was his second jump; the first occurred at the 1929 National Air Races during an off-airport practice session when the upper wing of a Curtiss Hawk failed.

In the wake of two accidents both the first *Shell Mystery Ship* and the Doolittle racer were gone, but not forgotten. Doolittle and Haizlip were soon flying other custom-built airplanes owned by the company, including the retractable-gear *Shell Lightning* Lockheed *"Orion."*

After leading the raid on Tokyo in 1942, for which he was awarded the Medal of Honor, Doolittle was promoted to Major General and given command of the Twelfth Air Force in North Africa, the Fifteenth Air Force in Italy during 1943, and in 1944-1945 as a Lieutenant General he led the most powerful air armada the world had ever seen—the United States Army's mighty Eighth Air Force that reduced Hitler's Third Reich to rubble.

When peace returned he became Director of the Shell Oil Company for many years and also served as an executive with a number of aviation and space corporations. Doolittle later entered the insurance business and was a director with Mutual of Omaha and other companies. He died September 27, 1993 at the age of 96 and was buried in the Arlington National Cemetery. He was one of America's greatest heroes of World War Two and one of the nation's prestigious aviators.

James Haizlip also had a distinguished career in aviation. In addition to his accomplishments flying the *Shell Mystery Ship* in 1930-1931, he won the 1932 Bendix Trophy Race flying a Wedell-Williams racer, completing the race in 8 hours, 19 minutes, 45 seconds. He later took off from Cleveland, Ohio for New York City and set a transcontinental record of 10 hours, 19 minutes. During the 1932 Cleveland races Haizlip won the event for engines with a displacement of 1,000 cubic inches and placed fourth in the Thompson Trophy Race.

Although he continued to fly for the Shell Oil Company, in April 1936 Haizlip took delivery of a new Beechcraft Model B17R (c/n 71, NC15816) in preparation for a trip to Europe where he planned to reside temporarily with his wife Mary and their son, James Jr. As a result of discussions with Walter Beech, Haizlip had agreed to act as an agent for the Beech Aircraft Company in Europe. With Beech's approval he had the B17R dismantled and loaded aboard the giant German Zeppelin *"Von Hindenburg"* for transport across the Atlantic Ocean. On May 21 the airship departed from Lakehurst, New Jersey on its second return flight to Europe.

Haizlip resigned from the Shell Oil Company in 1937 and served in the United States Army Air Forces in World War Two. During that conflict he flew more than 2,000 hours as a pilot assigned to the Special Projects Unit of the Air Transport Command. He remained in the service until 1947. [7]

After the war ended he and his wife Mary eventually relocated to California where he worked in the aerospace industry. The couple lived for many years in Pacific Palisades, California, where Jim eventually retired and Mary Haizlip became involved in the real estate business. James Haizlip died in 1983 at the age of 87. [8]

CHAPTER FIVE NOTES

1. The raid caused infinitesimal damage to Japan's war machine but was a significant blow to its military hierarchy. Americans, however, were thrilled by the attack and it served to bolster sagging morale following the Pearl Harbor disaster of December 7, 1941.
2. Letter to the author from James G. Haizlip dated August 29, 1983.
3. Bowers, Peter M.; *"Curtiss Aircraft - 1907-1947,"* Putnam & Company, Ltd., London, 1979.
4. Schmid, S.H., and Weaver, Truman C.; *"The Golden Age of Air Racing,"* Experimental Aircraft Association Foundation, 1991.
5. Doolittle did not contest the fine and admitted that the collision was his fault.
6. Letter to the author from James G. Haizlip dated August 29, 1983.
7. *"An Encyclopedia Of Aviation Greats;"* Buckeye Aviation Book Company, Reynoldsburg, Ohio, 1990.
8. *"The Aeroplane;"* September 15, 1973, p. 347, no. 1373, London; and the *"Palisadian Post"* newspaper, Pacific Palisades, California, March 26, 1987.

"Texaco 13"

R-2004 - NR1313

Of the five Type R monoplanes built, The Texas Company's NR1313 achieved more notoriety, set more records, and was more successful than any of her four sister ships. The airplane was purchased by The Texas Company to serve as a flying laboratory and to promote petroleum products and capture the public's imagination and admiration through the medium of speed.

Dubbed *"Texaco 13,"* the monoplane was ordered in May 1930 and built for the express use of the company's superintendent of aviation, Frank Monroe Hawks. Perhaps to a greater degree than his aviation contemporaries, Hawks seemed obsessed with speed and its value to commercial flying. Frank's 1929 transcontinental flight in the Lockheed Air Express (see sidebar) had whetted his appetite for setting records and by 1930 he was anxious to find an airplane worthy of his aspirations. Hawks was well aware of the Travel Air Type R and its performance capabilities, and Doug Davis's victory at the 1929 National Air Races had helped to convince Frank that the single-seat ship was the instrument he needed in his quest to rewrite the record books.

With The Texas Company's support he contacted Travel Air and placed an order for a Type R to be built to his specifications for long-distance flying. As part of the purchase, Hawks planned to have a 400-hp., high-compression Wright J-6-9 radial engine installed in the ship. But what the airplane needed most was something unusual that would attract the public's fancy. Frank soon found the answer. On May 14, 1930 he wrote to Clarence M. Young, Assistant Secretary for Aeronautics at the Department of Commerce in Washington, D.C. and asked for a special favor: "I am purchasing a new airplane for my own use and it so happens that it is the thirteenth of our fleet. Consequently, it will carry the (designation) *"Texaco 13."* It occurs to me that it would be rather unique and advantageous to the Company if I could possibly secure a license number of either 13 or 1313." Hawks was unconcerned about superstitions associated with the number "13."

His unusual request was soon answered by the Aeronautics Branch. Gilbert G. Budwig, director of air regulation, telephoned Hawks and arranged to have license NR1313 assigned to the new Travel Air. He also

The Texas Company's Type R, NR1313 (R-2004) was built to specifications stipulated by Frank M. Hawks and was completed in June 1930. Hawks planned to use the airplane chiefly for record flights, not closed-course pylon racing. It was photographed at Travel Air Field soon after Clarence Clark had made the initial test flights. ***(Smithsonian Institution negative number A-43524D)***

Rear view of NR1313 shows standard wing installation, placement of Texaco logo on the left wing, and widened fuselage that accommodated three fuel tanks with a capacity of 109 gallons. Access panel for the fuel tank caps is visible on the top of the fuselage directly aft of the cowling. The three-piece cockpit enclosure was designed by Hawks, who closely monitored construction of the airplane. ***(Raytheon Aircraft Company via Mike Madewell)***

Frank Hawks cast a casual glance at the camera as he prepared to fly Texaco 13. The upper canopy has been slid aft and the pilot's seat has been elevated to the takeoff position to improve forward visibility. Note the raised, reinforced section of the fuselage aft of the cockpit that protected the pilot if the airplane crashed and turned over on its back. ***(Kenneth D. Wilson)***

forwarded the number to the factory in Wichita. But there was a catch. On May 16 Clarence Young wrote to Hawks politely reminding him that the "R" license was issued only as a temporary measure and was "...premised upon the fact that you propose to use the ship for racing purposes only. However, it is a commercial use within the meaning of the Air Commerce Act and should carry a "C" license. The "R" will serve your immediate purpose and permit the Travel Air Company to submit material in the ordinary manner for an A.T.C." (Approved Type Certificate).

The Aeronautics Branch did not normally issue a "Restricted" license for racing because it fell within the definition of a commercial activity for the purpose of earning money. The "R" category was reserved for "...specific industrial applications such as aerial photography, crop dusting, sky writing and similar operations," Gilbert Budwig wrote to Walter H. Beech on May 27,

In-flight photographs of the five "Mystery Ships" were rarely made, but Hawks was captured on film flying high above the Kansas Plains during an acceptance flight early in July 1930. The pilot's seat is raised and the canopy slid forward to increase ventilation in the cockpit. A special vent directed fresh air to the pilot, but it was often inadequate for flight on hot days. ***(Raytheon Aircraft Company)*** ▼

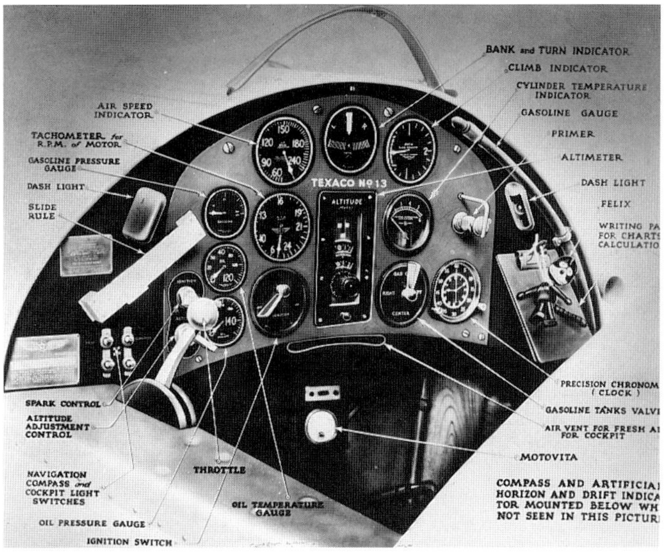

The instrument panel and cockpit of NR1313 were equipped with special instrumentation to help Hawks during his cross-country flights. Note "Felix the Cat" toy and writing pad as well as "Moto-vita" gauge for adjusting the fuel/air mixture in flight. ***(Smithsonian Institution negative number 81-3250)***

1930. In return for issuing the "R" license to Hawks, the Aeronautics Branch was expecting the Travel Air Company to quickly submit the necessary paperwork to have the Type R approved for a commercial ("C") license. That was, however, not possible and Travel Air's chief engineer Herbert Rawdon had informed Walter Beech of that fact.

The *Texaco 13* would not qualify for an Approved Type Certificate as stipulated in Aeronautics Bulletin No. 7A because it had a landing speed that was in excess of 60 mph., a payload of less than 200 pounds, and was fitted with an engine that operated at more than 105% of the rated R.P.M. At Beech's request, Rawdon wrote to Budwig late in May and stated the situation bluntly: "It is impossible for us to get a "C" rating on Travel Air "Mystery Ships." This design, of course, is sufficiently

The Texaco 13 was damaged when Hawks crashed at Travel Air Field in July 1930. The airplane was repaired and repainted within a few weeks of the accident while Hawks recuperated in a Wichita hospital. ***(John Sunyak)***

strong structurally but on account of the present limitations (in Aeronautics Bulletin 7A) it will be impractical and of little use for us to compile the necessary engineering data required for an A.T.C."

As a result, data were never submitted by Travel Air but *Texaco 13* wore (N)R1313 throughout its career. Walter Beech, who supported the need for regulations but abhorred the micromanagement of bureaucrats assigned to administer them, tried a different approach to resolve the issue. In July 1930 he wrote to Budwig with a request: "Is it possible to make concession and give Travel Air Mystery Ship numbers 2002, 2003, and 2004 "C" licenses. We have discontinued the manufacture of this model, but we wish to get a "C" license for these three ships as a special favor to our clients as the renewing of the "R" license becomes somewhat of a bother." Budwig swiftly and steadfastly refused, but assured Beech that the Aeronautics Branch would continue to cooperate and issue "R" licenses to the airplanes. [1]

Publicity photograph released by Texaco with Frank Hawks and NR1313 after the airplane was repaired. Note the hole in the right side of the cowling for a handcrank, which was inserted to engage the intertia starter. ***(Texaco)***

Another in-flight photograph of NR1313. Powered by the high-compression Wright J-6-9 engine the airplane's maximum speed was about 250 mph. Note that Hawks has elevated his seat and slid the upper canopy forward to improve forward vision and ventilate the cockpit. ***(Peter M. Bowers)***

In his 1931 book entitled "Speed," Hawks wrote that the "Purchase of the Mystery Ship necessitated my presence at the Travel Air factory in Wichita a greater part of the time during its construction so that I could supervise the work, be on hand to make decisions when various details came up that had to be settled, and generally to see that the plane was built to suit me in every particular." Late in June R-2004 was completed and on July 3, 1930 The Texas Company paid $22,706 for the airplane.

It originally was powered by a high-compression Wright J-6-9 radial engine (serial number 10777) equipped with a Hamilton Standard steel propeller (serial number 12509). The engine was the same J-6-9 that was later modified and installed on Shell Oil Company's NR482N for the 1930 Thompson Trophy Race. In addition, the airplane featured three fuel tanks in the fuselage that provided a total capacity of 109 gallons.

The fuel tanks of Texaco 13 are being replenished during one of two transcontinental speed dashes made by Hawks between California and New York City early in August 1930. Note the venturi tube mounted on the left side of the fuselage that operated a gyroscope in one of the flight instruments. A second venturi was mounted on the right side and operated another gyroscopic instrument. ***(Raytheon Aircraft Company)***

According to Texaco records, a set of wings for closed-course pylon racing were built that had a span of 11.6 feet per panel and a chord of 60 inches, but these were not installed initially. As delivered to Hawks *Texaco 13* was equipped with standard wings having a span of 30 feet, 6 inches. The upper fuselage spine aft of the cockpit was elevated to protect the pilot if the airplane went over on its back in a landing accident. Aeronautics Branch inspector Walter L. Pike carefully perused the airplane before approving it and issuing a racing license that was valid until October 15, 1930. [2]

Hawks had the airplane finished in a striking paint scheme that drew attention to the monoplane. When Hawks accepted the airplane on July 5 NR1313's wings were painted Stearman vermilion with the license and registration number painted in cream with blue trim. The fuselage was a light gray, and the four-piece, NACA-type cowling was painted vermilion as was the vertical stabilizer. These were trimmed in vermilion. The horizontal stabilizers were vermilion overall, as was the landing gear that featured cream flashing trimmed in blue. The Texaco logo was painted on the wings and vertical stabilizer in red, green, white, and black. *"Texaco 13"* was painted on the sides of the fuselage in vermilion.

The cockpit was sophisticated and well equipped for a high-performance airplane of the era. In addition to the standard complement of engine and flight instruments for "blind flying," Hawks installed a sliderule for making in-flight calculations of speed and wind drift, a precision chronometer, and a relatively new device called a "Moto-vita." Developed in the late 1920s by Miller Reese Hutchison, it was a forerunner of modern exhaust gas temperature sensing systems. The unit incorporated a Wheatstone bridge electrical circuit that used platinum and nickel wires whose electrical resistance was the same at equal temperatures. When the engine was operating these wires responded to changes in the composition of gases in the exhaust stream (typically nitrogen, carbon monoxide and carbon dioxide) that created an imbalance in their resistance values. The small electric current that was generated by that process drove a meter in the cockpit to indicate the percentage of gases in the exhaust. As a result, the pilot could adjust the mixture to obtain optimum combustion efficiency for a given power setting. Hawks would use the Moto-vita to good advantage on his many record-setting flights. [3]

The cockpit also included one other interesting piece of equipment—a toy replica of the cartoon character "Felix" the cat, complete with flying helmet and goggles. Hawks hung the toy on a small clipboard he used for scribbling notes and mathematical calculations. A dry cell battery powered lights that illuminated the instrument panel for night flying, including the sliderule and a tube-like, glass sight gauge that indicated approximate fuel quantity.

Hawks, who had visited the factory when the airplane was being built, was satisfied that the new machine met his requirements and was anxious to fly it. Test pilot Clarence Clark already had flown the ship in preparation for Frank's evaluations. On a hot July 5 he strapped himself into *Texaco 13* and prepared for his first flight in the airplane. Frank took off from Travel Air Field and returned less than one hour later. Being unfamiliar with the ship, on final approach to the runway he came in too fast and had to abort the landing and go around for another attempt, which was successful.

Frank was pleased with the airplane's performance but became increasingly concerned about the high oil temperatures he had observed on the flight. The airplane had been equipped with oil coolers installed on each side of the stub wing. These were necessary to keep oil temperatures within limits during high-speed, long-distance flying. On July 10 Hawks made another flight and the temperature problem returned. Adjustments were made to the inter-cylinder baffles but the condition persisted. During another test flight on July 11 the engine failed as Hawks was maneuvering for final approach north of Travel Air Field.

According to Frank's description of the incident, he decided the best place to set the ship down was a field full of dead trees and stumps. Instinctively he moved the stick forward in an effort to maintain airspeed and stretch the glide as far as he could. But as the airplane descended Hawks realized that he was flying too fast to safely land between the stumps. Reacting quickly and with no time to waste, he redirected the Type R toward the runway but a line of telephone poles festooned with wires stood in his path. The bold maneuver almost worked, but the Travel Air's landing gear snagged the wires and the ship slammed into the ground nose-first at about 80 mph., sliding a short distance before turning over on its back.

Fortunately, there was no fire. A small group of people were on the scene almost immediately and extracted an unconscious Hawks from the cockpit. He was uninjured except for a few bruises and a cut on the back of his right hand, inflicted when one of the wires that had snapped under the ship's weight flew back and struck him in the cockpit. The airplane, however, was seriously damaged but could be repaired. The propeller, engine, and cowling were smashed and the vertical stabilizer was crushed almost flat. Hawks was taken to a Wichita hospital where he was treated for head injuries and confined to bed for a few days rest before being released. Frank wanted the ship rebuilt as soon as possible and decided to take advantage of the situation to make some changes to the aircraft. Work began almost immediately after the accident.

Although Travel Air and Department of Commerce records are vague, according to Texaco records during the rebuilding process Hawks installed a 465 hp. Wright J-6-9 engine (serial number 13434) featuring a higher supercharger ratio and other modifications. With a maximum gross weight of nearly 2,800 pounds *Texaco 13* was heavier than any of the other four Type R ships built by Travel Air. Hawks may have determined that additional horsepower was necessary to maintain the monoplane's high performance capability.

The bent propeller blades were straightened and repaired by Hamilton Standard and shipped back to Travel Air. A new propeller for the more powerful engine had been ordered from the propeller company but was not ready. Steel tubing in the forward fuselage section was replaced, the vertical stabilizer and rudder were

Rear, quarter view of NR1313 illustrates the short-span wing installation. The racing wings normally were not wired for navigation lights. Note that the navigation lights had been removed from the vertical stabilizer. ***(Raytheon Aircraft Company)***

Frank Hawks poses with NR1313 at Travel Air Field. Between August 1930 and April 1932 he flew the airplane more than 125,000 miles on two continents demonstrating the speed potential of commercial airplanes. ***(Carl Burnham)***

straightened and repaired, and new struts were installed on the right landing gear and other repairs made to both gear assemblies. In addition, the plywood covering on the wing leading edges was repaired as required. To complete the transformation of *Texaco 13,* the airplane was painted in a new scheme that consisted of broad, red and white stripes on each side of the fuselage that extended from the cowling to the tail. [4]

The repairs were completed on July 23 and were approved by inspector Walter Pike on July 26, 1930. After a brief series of flights by Clarence Clark to check the overall rig and handling of the aircraft, Hawks flew it and pronounced it satisfactory. According to Hawks, maximum speed was in excess of 240 mph. and he found the airplane handled well on final approach at 70 mph.

The recent delays caused by the accident and rebuilding program had frustrated Hawks and interrupted his timetable for setting records with the new ship. After the new propeller was installed Frank flew *Texaco 13* from Wichita to Roosevelt Field near New York City, where

Frank Hawks shakes hands with a woman at Quebec Airport in Canada after returning from a tour of Europe in June 1931. Note the oil streak on the lower part of the cowling, and minimal exhaust stains on the fuselage. ***(John Sunyak)***

Front, quarter view of the airplane shows struts and landing gear fairings that had not been painted when the photograph was made. A small, vertical support brace was installed between the wood vibration damper rod and the upper surface of the wing. ***(Raytheon Aircraft Company)***

the airplane was christened by Mrs. Hawks and Texaco officials. A few days later Hawks and NR1313 were poised for an assault on the east-to-west transcontinental speed record established earlier that year by the flamboyant Roscoe Turner.

At 6 A.M. on August 6, 1930 Hawks and *Texaco 13* took off from Curtiss Airport, Valley Stream, N.Y. and headed west toward California. Despite inclement flying weather and incessant headwinds, Hawks landed in Los Angeles 14 hours, 30 minutes, 43 seconds later, establishing an east-to-west record and beating Turner's time by more than three hours. Only one week later man and machine were ready to attempt a west-to-east flight to capture that record, which had been set in May by Charles Lindbergh in a Lockheed *Sirius* monoplane. According to Hawks, in preparation for the flight the three fuel tanks were loaded with 80 gallons of Texaco aviation gasoline and the oil tank was serviced with 10 gallons of Texaco No. 2 lubricating oil.

Hawks took off from Glendale Airport in California at 2:16 A.M. Pacific Time. He crossed over San Bernardino Range through a pass at 4,000 feet and followed a northeasterly route until reaching the Mojave Desert. His speed up to that time was greater than three miles a minute. It was still dark when he passed the National Forest in Arizona, and he recalled later that "I could see below me the lights on freight trains plugging along at about twenty miles per hour." The sun began to rise as he passed Flagstaff where Hawks hoped to encounter a favorable tail wind, but to no avail. After adding one gallon of oil and refueling NR1313 at Albuquerque, New Mexico Hawks climbed to 10,000 feet and received an extra push toward New York from a 34-mph. tail wind that continued for the next 300 miles.

The airplane was refueled again at Wichita. At that point the flight was at its halfway mark and Hawks had been flying for only five hours, 45 minutes since leaving the West Coast. He departed Wichita and pressed eastward toward St. Louis, Missouri and landed again for 80 gallons of fuel at Indianapolis, Indiana where four gallons of oil also were added. The Wright engine continued to roar along as *Texaco 13* passed over Ohio and into Pennsylvania at a rate of nearly four miles per minute. Soon the "...tall towers of Manhattan became visible.

In January 1931 Hawks flew to the United States Naval Base at Anacostia, Maryland and offered the use of Texaco 13 to David S. Ingalls, Secretary of the Navy for Aeronautics. The airplane remained at Anacostia during February while Hawks and humorist Will Rogers toured the nation in a Curtiss "Helldiver" biplane. The photograph, dated February 15, 1931, shows the racer with a map of the transcontinental records set in 1930. Note pan under the engine to catch oil seepage. **(W.O. Collier Collection)**

Over New York I started to dive straight for the airport at Valley Stream, which I circled before coming in for a final glide and a fishtail landing at seventy miles per hour," he later recalled in an article published in *Aero Digest* magazine. He landed at 6:41 P.M. Eastern Time.

Hawks landed the airplane 12 hours, 25 minutes, 3 seconds after leaving Los Angeles—a record that superceded Charles Lindbergh's feat by more than two

Travel Air employee Theodore "Ted" Cochran sits in the cockpit of NR1313 during one of the airplane's visits to the factory for modifications and maintenance. Cochran, who was a skilled woodworker, was one of the team members who built the Type R monoplanes. **(Raytheon Aircraft Company)**

The Texaco 13 is carefully hoisted aboard the S.S. Europa for a voyage to Europe in April 1931. The cockpit and engine sections have been covered with tarpaulins to minimize salt intrusion during the trip. Gust locks have been installed on all of the control surfaces. **(Smithsonian Institution negative number 83-3560)**

America's "King of Speed" poses with NR1313 after arriving in Europe. ***(Raytheon Aircraft Company via John W. Underwood)***

hours. Frank's actual time in the air was 11 hours, 40 minutes, 3 seconds after subtracting 45 minutes for three fuel stops. Groundspeed for the 2,510-mile trek was, "never less than 200 miles an hour" and on some stretches reached 260 miles per hour, according to Hawks. He ran the engine at 2,150 RPM. "all the way" and noted that oil temperature did not exceed 65 degrees Centigrade, according to *Aero Digest* magazine. These initial transcontinental flights, however, were only the first of what would become a string of more than 200 inter-city records Hawks established with *Texaco 13* between August 1930 and April 1932. [5]

In a story published in the *"The American Magazine"* Hawks revealed one of the chief reasons he selected the Travel Air Type R. During his 1929 transcontinental flights in the Lockheed Air Express he suddenly realized he was not only competing against the clock, but also against the sun. "When you fly from east to west you are really racing the sun. The sun also is traveling east to west. When I started out for Los Angeles that morning, the sun was rising at my back, well behind me. But it was making eight hundred miles an hour. Before I was half way across it had caught me and was over my head. Then it began to get in front of me. I was chasing it. When it vanished over the shoulder of the world I was still in the air. When I landed in Los Angeles it was dark. It irked me that the sun should leave me so far behind."

When Hawks flew from New York to California in the new *Texaco 13* he "managed to stick to his flaming rival well enough to land in Los Angeles when it was still in the sky, nine minutes and a fraction before six o'clock," the magazine article stated. Such was the difference in speed between the Lockheed and the Travel Air. Later Hawks said speed flying "promises astonishing possibilities in future travel. It won't be so long now before we'll be able to catch up with the sun enough to have breakfast in New York and lunch in Los Angeles. And maybe that won't be all. When we learn to go away up high, in regions where the atmosphere is thin and resistance almost nil, marvelous speeds will result. We'll keep pace with the sun. We'll be having lunch in New York and in Los Angeles lunch again!" With the dawn of transcontinental jet age in 1959, Frank's prophecies

A cameraman in Europe photographed Hawks in the cockpit wearing his fur-lined flying suit. Note the venturi tube, and the fabric above the aft cabane strut that was starting to separate from the underlying sheet metal. ***(Raytheon Aircraft Company via John W. Underwood)***

began to be fulfilled.

According to air racing historian Truman C. Weaver, following the coast-to-coast flights Hawks flew to Wichita in mid-August 1930 and had the set of short-span wings installed on the monoplane for the upcoming National Air Races in Chicago, Illinois. Unfortunately, *Texaco 13's* performance at the races was less than auspicious. Hawks, despite his lack of experience in high-speed, closed-course pylon racing, had entered the airplane in the Thompson Trophy Race and was confident that the airplane would be competitive. [6]

Assigned race number "28," Hawks and NR1313 were doing well at the beginning of the event but he was soon passed on the third or fourth lap by James G. Haizlip in the *Shell Mystery Ship.* Soon after Haizlip passed *Texaco 13* Hawks was forced to retire from the race because of fuel flow problems. He landed without incident and returned to the Texaco parking area in front of the grandstands. It was later determined that improper venting of the fuel tanks had interrupted fuel flow, causing the engine to lose power. It should be stated that *Texaco 13's* fuel tank caps were located inside of the fuselage, aft of the engine firewall and were not directly exposed to the airflow during flight.

After the races concluded Hawks returned to setting records. On September 9, 1930 he flew non-stop from New York to Ford Airport near Detroit, Michigan in 3 hours, 3 minutes, eclipsing the previous mark of 4 hours, 15 minutes set by Charles Lindbergh in 1928. During the next 21 days he unleashed an unprecedented assault on inter-city speed records that dazzled the public and earned Texaco a barrage of attention from the press. On October 1 Hawks flew from Detroit to New York 2 hours, 41 minutes followed on October 7 by a 190-mile flight from Boston to Valley Stream Airport in New York in a mere 53 minutes. The next day he took off from William Penn Airport near Philadelphia, Pennsylvania and 20 minutes later landed in New York City to deliver photographs of the last baseball game of the 1930 World Series. On October 21 Hawks guided NR1313 from New York to Washington, D.C. in 71 minutes and returned to

FRANK MONROE HAWKS

"The King Of Speed"

Frank M. Hawks and his exploits with *Texaco 13* made him one of America's best-known aviators of the early 1930s, but he was not a racing pilot in the pure sense of the term. Instead, his passion was selling the concept of speed and the commercial advantages of the airplane to the public.

Born in Marshalltown, Iowa on March 28, 1897 Frank eventually moved with his family to Minnesota and to Long Beach, California where he attended high school. During his school years he worked part-time for the Christofferson brothers, who gave airplanes rides and flight instruction at an airport in the Long Beach area. Under the brothers' tutelage Hawks learned the rudiments of flying an airplane. After graduating from high school in 1916 he entered the University of California and was active in varsity sports. Standing five feet, 6 inches tall and athletic in form, Hawks enjoyed track and field events as well as football, but described himself as a "rotten golfer." He shunned card games and dancing.

When the United States became involved in World War One, the 20 year-old Hawks joined the United States Army. In July 1917 he was assigned to the U.S. School of Military Aeronautics in Berkeley, California and in September received flight instruction at Love Field in Dallas, Texas and also attended gunnery school in Fort Worth, Texas. In November he was assigned to the Army's Pursuit School at Lake Charles, Louisiana before attending instructor's school at Brooks Field, Texas in April, where he remained until being discharged in March 1919 with the rank of captain. [8]

For the next five years Frank barnstormed southern California and gained further experience in the air by flying oil payrolls in remote sections of Mexico. During one of his flights to Tampico, Mexico he met an amateur actress from Texas named Edith Bowie. They were married a year later in Houston, Texas where Hawks had found a job dusting crops with arsenic to fight insect infestations.

In 1927 Edith Bowie Hawks helped to finance Frank's purchase of a Mahoney Ryan B-1 *Brougham* cabin monoplane that her husband named "The Gold Bug" because of its color. Next, Frank persuaded the makers of Maxwell House coffee to help finance his participation in the 1927 National Air Tour for the Edsel B. Ford Trophy, held June 27-July 12, in return for advertising the coffee. Frank, accompanied by Edith and some friends, flew the *Brougham* more than 4,000 miles and completed the aerial marathon.

Following the Ford Tour Frank's career began to climb. In 1928 he was hired by The Texas Company to fly *"Texaco 1,"* a Ford Trimotor. His big break, however, came in February 1929 when he ferried a new Lockheed *"Air Express"* from Los Angeles to New York in 18 hours, 21 minutes, 59 seconds. Lockheed employee Oscar Grubb accompanied Hawks to pour cans of fuel into the tanks as the journey progressed. The flight erased the record set in August 1928 by pilot Arthur C. Goebel and Harry Tucker in the Lockheed Vega *"Yankee Doodle."* They were, however, the first to fly nonstop across the United States from west to east, and the first to complete a coast-to-coast flight in fewer than 24 hours.

In the wake of national publicity generated by Frank's west-to-east flight in the *Air Express,* The Texas Company acquired the airplane and dubbed it the *"Texaco 5."* Flying solo, on June 27-28, 1929 Hawks guided the Lockheed from New York to Los Angeles in a record 19 hours, and after an eight-hour pause flew the parasol monoplane back to New York to set another record. In addition to accomplishing that significant feat in only two days, Hawks became the first person to fly solo coast-to-coast in both directions.

Frank was a gregarious individual whose friendly smile was backed up by a mouthful of gleaming white teeth. He had a penchant for spinning lively tales about his aeronautical exploits and his repertoire seemed endless. Frank was especially fond of recounting a record he had made in 1921 flying a decrepit Standard J-1 biplane. He and a wing-walker by the name of Wesley May were aloft in the Standard over Long Beach, California to rendezvous with another gypsy pilot and barnstormer named Earl Daugherty who was flying a Curtiss *"Jenny."* In an attempt to refuel the old JN-4 in flight, the two ships were flown in tight formation and May was able to make a transfer to Daugherty's machine and pour a five-gallon can of gasoline into the Jenny's fuel tank. As a result, Hawks claimed he was the first person to pilot an aerial refueling tanker.

Hawks prepares to start the Wright J-6-9 engine in NR1313. A new, larger venturi had been installed on the left side of the fuselage under the front brace wire. ***(Smithsonian Institution negative number A-44744)***

Frank also caught the public's attention in April 1930 when he piloted an Eaglet glider from San Diego, California to New York City. The aircraft was towed behind *"Texaco 7,"* a Waco biplane flown by J.D. "Duke" Jernigan. The transcontinental flight required eight days and 35 hours in the air. Another 10 hours were flown during soaring performances at stops along the route. The glider was later donated to the Smithsonian Institution.

After the crash of *Texaco 13* in April 1932 Hawks recovered from his injuries and returned to flying for The Texas Company that summer. He also had resigned his reserve commission in the Army Air Corps to become a lieutenant commander in the U.S. Navy. But instead of his faithful Travel Air, Hawks would pilot a sleek, all-metal monoplane known as the Northrop *"Gamma"* and dubbed *"Sky Chief"* by Texaco. Designed by Jack Northrop, the *Gamma* was a fast machine and Hawks wasted no time in demonstrating its speed to the public. In June 1933 he established a transcontinental record by flying from Los Angeles to New York City's Floyd Bennett Field in 13 hours, 30 minutes.

Christened "Meteor Man" by the press, Hawks set more inter-city records with the big Northrop monoplane including a trans-Canada flight between Vancouver, British Columbia and Quebec City. Although Hawks set additional records with the graceful *Gamma,* in May 1934 he resigned from The Texas Company to pursue other interests. Among these was his quest to acquire a 300-mph. airplane, which remained unfulfilled until 1936. By June of that year Frank and engineer Howell "Pete" Miller were conducting wind tunnel tests of a new speed ship, the HM-1, and construction was scheduled to be completed in September.

The retractable-gear HM-1 was designed chiefly by Miller, who also contributed to the design of the famed *"Gee Bee"* racers built by the Granville Brothers in Springfield, Massachusetts. Frank planned to use the monoplane for advertising, racing, and long-distance and transcontinental flying. In keeping with his penchant for thirteens, Hawks requested and was assigned registration number "1313" for his new airplane.

Powered by 1,150 hp. Pratt & Whitney R-1830-SB-G radial engine, on April 3, 1937 Hawks flew the HM-1 from Washington, D.C. to LaGuardia Airport in New York in 48 minutes. On April 13 he made a trip from Hartford, Connecticut to Miami, Florida to New York in 9 hours, 16 minutes. After making a hard landing that damaged the HM-1, on May 2, 1938 Hawks sold the airplane to Tri-American Aviation in New York. The airplane later was destroyed when a wing failed during an aerobatic demonstration at East Hartford, Connecticut on October 3, 1938. The pilot, Earl Ortman, parachuted to safety.

Discouraged after the crash of the HM-1, Frank Hawks abandoned speed flying but his interest in aviation remained strong. Known nationwide as the "King of Speed," it was ironic when on August 23, 1938 Hawks was fatally injured in a slow, unconventional airplane known as the Gwinn *Aircar* that had been developed by Joseph Marr Gwinn. Frank and a passenger had departed in the odd-looking aircraft from a small field near Buffalo, New York when it struck telephone lines and crashed. The passenger was killed, but Hawks clung to life for a few more hours before dying in a Buffalo hospital. He had been flying for 21 years when the accident occurred. [9]

New York later the same day making the 225-mile trip in only 63 minutes. The series of inter-city speed dashes had been so successful that some people were soon chanting, "Don't send it by telegraph, send it by Hawks!"

The Texaco 13 after landing at an aerodrome in England. Texaco crews based in Europe assisted Hawks and serviced the airplane during his visit. A company fuel truck is visible behind the airplane, and a group of spectators had gathered on the rooftop of the building to watch. The heavy exhaust stains on the forward fuselage indicate that the airplane had been flying for an extended period of time that day. ***(Raytheon Aircraft Company via John W. Underwood)***

Frank's chief reason for visiting the nation's capital was to demonstrate *Texaco 13* to United States Navy Rear Admiral William A. Moffett and David S. Ingalls, Assistant Secretary of the Navy for Aeronautics. Later, in January 1931 Ingalls allowed Hawks and humorist Will Rogers to fly his Navy Curtiss *"Helldiver"* on a month-long trek around the mid-western states to raise money for relief of drought-stricken farmers in the region. In response, Frank placed *Texaco 13* at the Ingall's disposal and the airplane was hangered at Naval Air Station Anacostia, near Washington, D.C. Hawks, Rogers and famed pilot C. S. "Casey" Jones flew from New York to Washington in a Cessna cabin monoplane to begin their journey in the Curtiss biplane that would take them to 57 cities and towns. Ingalls, who was an accomplished aviator and a World War One pilot, flew back to New York with Jones in the Cessna. He later flew *Texaco 13* back to Anacostia.

Hawks taxies out for takeoff in England. Only small exhaust stains are visible on NR1313, perhaps indicating that he was about make the first flight of the day. The monoplane was fitted with standard wings for most of its career, including the European tour. ***(Raytheon Aircraft Company via John W. Underwood)***

During his attack on inter-city records in the autumn of 1930, Frank had told the press that the flights were made to prove a point and were merely a precursor to greater things to come: "The transcontinental flights and the minor trips were made to demonstrate the feasibility of *Texaco 13* as a high-speed ship. I now propose to fly to Havana in an effort to prove further that this plane and other fast aircraft can be profitably adapted to commercial light express work." The distance between Havana, Cuba and New York was about 1,400 miles. Hawks had *Texaco 13* carefully prepared for the flight south and on November 6 he departed New York and landed at Jacksonville, Florida before taking off 23 minutes later for Miami, Florida where he refueled and flew on to Havana. Total flight time was 8 hours, 38 minutes, or 9 hours, 21 minutes by the clock.

Hawks planned on resting for a day and flying back to New York on November 8, but bad weather intervened. The skies were beckoning him on November 9 and he departed Havana and landed in Miami 1 hour, 33 minutes later. He flew to Charlotte, North Carolina in 3 hours, 40 minutes and completed the flight to Roosevelt Field in New York in 2 hours, 50 minutes for an elapsed time of 8 hours, 44 minutes. Hawks estimated that his groundspeed for most of the flight was about 195 mph. "I have not been punishing my plane, but instead have utilized its high cruising speed to establish these marks," he told *Aero Digest* magazine after the flight.

In January 1931 Hawks flew from Forth Worth, Texas to New York City in eight hours, 30 minutes followed in February by a flight between Boston and New York City in 55 minutes. In March he made the same trip in 52 minutes. Early in 1931, however, Hawks began planning his most ambitious journey to date with *Texaco 13.* He planned to take the monoplane by ship across the

Another view of Texaco 13 departing England for a major European city. Hawks already had his flying goggles in position. Men at each wing tip were stationed there chiefly to avoid hitting obstacles or other aircraft, not to help Hawks maneuver the airplane on the ground. The ship was equipped with a differential, toe-operated brake system that improved maneuverability on the ground. **(Raytheon Aircraft Company via John W. Underwood)**

Atlantic Ocean and, while conducting Texaco business on the Continent, casually demonstrate the airplane's performance to the European community. The trip also would allow Hawks to consult with foreign airplane manufacturers and pilots as well as experience European airports and air transportation.

With the distances between nations and major cities generally much shorter than in the United States, Hawks looked forward to touring Western Europe by air. He contacted Clarence M. Young, who was then serving as Assistant Secretary of Commerce, and explained his intentions along with seeking permission to overfly the

Frank Hawks flashed his famous smile for the camera after arriving at Tempelhof Airport in Berlin, Germany from London in two hours, 57 minutes on May 12, 1931. The nearly 200-mph. cruising speed of Texaco 13 did not impress the Europeans as much the airplane's ability to achieve that speed day after day without mechanical failures. **(Raytheon Aircraft Company via John W. Underwood)**

following countries: Belgium, France, Germany, England, Ireland, Scotland, Sweden, Denmark, Italy, Norway, Czechoslovakia, Austria, Hungary, and Rumania. He also proposed to visit Russia if arrangements could be made.

Young discussed the request with Assistant Secretary of State Francis White in mid-March 1931 and the State Department secured the necessary authorizations through diplomatic channels. On April 1, 1931 Hawks and his wife were aboard the steamship S.S. *Europa* after ensuring that *Texaco 13* was safely stowed, covered, and secured for the voyage (the aircraft was not disassembled). Hawks' faithful Texaco mechanic, Pete Clausen, had preceded him to France and was making arrangements to off-load the airplane at the port of Cherbourg.

During the five-day cruise to France both Hawks and his airplane often were the topic of conversation at dinner tables and during strolls around the deck. The most frequent question asked was whether Frank intended to break all of the inter-city speed records with NR1313. "You're taking your fast ship, aren't you? That seems very significant," the skeptics said. "No, a thousand times no," was Hawks' standard reply. He tried to explain that the airplane "has been a pal of mine for more than nine months now. We have traveled close to 10,000 miles at an average speed of 197 miles per hour. The motor has just been cruised at about half-throttle and is in perfect shape with barely 200 hours of flying time chalked against it. I have a genuine affection for my little ship and, I ask you, why shouldn't I take it along for transportation on the other side?" The skeptics, however, remained unconvinced of Frank's intentions.

The ocean liner arrived at Cherbourg on April 5 and by 5 A.M. Hawks and Clausen had the Type R removed from the *Europa* and made ready for flight. After receiving clearances from the French government he took off for Paris but had to deviate to skirt restricted areas of airspace. Including delays the flight required only 56 minutes at an average speed of 228 miles per hour. Later that day Hawks flew to Brussels, Belgium. While there he went to the main airport on a Sunday morning to observe

At the Amsterdam Airport on June 6, 1931 Texaco 13 was dwarfed by the prototype (registered D-2000) German Junkers G38 four-engine transport operated by Deutsche Lufthansa airlines and named "Deutschland" (it was destroyed in a crash at Dessau, Germany in 1936). The heavy rains had begun to wash off some of the exhaust stains on the Travel Air's fuselage. On May 12 Hawks flew from Berlin to Hamburg, Germany in 55 minutes. ***(Raytheon Aircraft Company via John W. Underwood)***

the flying activities—there were none! The airlines did not fly on Sundays and private flying was limited. With the airspace virtually all to himself, Hawks departed Brussels and spent the next two hours sightseeing along the coast, flying to Antwerp, and overflying some of the great battlefields of history. The next day he flew to Rotterdam, Holland before returning to Brussels on April 16. In his diary of the European trip he wrote that on the return flight *Texaco 13* traversed the 72-mile distance in 19 minutes 50 seconds. "There is a difference of time between these two cities; Rotterdam is 20 minutes faster than Brussels. I left the Dutch city at exactly 4 P.M. and arrived at the Belgian capital at the same time. If I had opened up my motor I would have arrived before I left."

On April 19 Frank flew to Croydon Aerodrome outside of London. He noted that the airport was "certainly a dandy" and that "there is no small amount of traffic coming and going at all hours of the day. When you consider that all air lines in Europe converge here, you can readily imagine that the airport must have sound traffic regulations." The next day he flew to Rome but bad weather delayed his return via Paris until April 23. He made the short flight from Le Bourget Airport in Paris to Heston Air Park in England in 59 minutes. In May Hawks and NR1313 accompanied the National Air Tour of France, after which he was asked to give a brief demonstration of the airplane. Frank wrote that "I don't do acrobatic flying any more, so I just flew an imaginary race with myself to give the people some idea of what 250 miles per hour looks like. I made a few turns around phantom pylons and came to a landing." The French were impressed by the demonstration.

On May 12 *Texaco 13* again was in the air headed for Berlin where it arrived 2 hours, 57 minutes later. According to Hawks, "I averaged well over 200 miles an hour all the way. The English Channel was enveloped in a dense fog over which I hopped. A nice day, a tailwind, and good visibility, much ceiling, and the sweetest-running motor you can imagine sitting behind." He received a warm reception at Tempelhof Airdrome and the next day was off to Hamburg before flying on to Linkoping and Stockholm, Sweden on May 16 to attend the International Aircraft Show. Hawks made a 900-mile sightseeing flight around Sweden and on May 22 completed the fastest trip of his tour by flying slightly more than 1,000 miles from Stockholm to Paris in 4 hours, 45 minutes. He credited a stiff tailwind for the relatively short time enroute. Upon Frank's return to the City of

After returning to the United States in June 1931 Hawks continued to set inter-city records with NR1313. A map of Europe adorned the fuselage along with a map of the United States. An antenna wire is visible between the cockpit and the vertical stabilizer. Note that the pitot tube had been bent upward and that stones had dented the engine cowling. ***(Peter M. Bowers)***

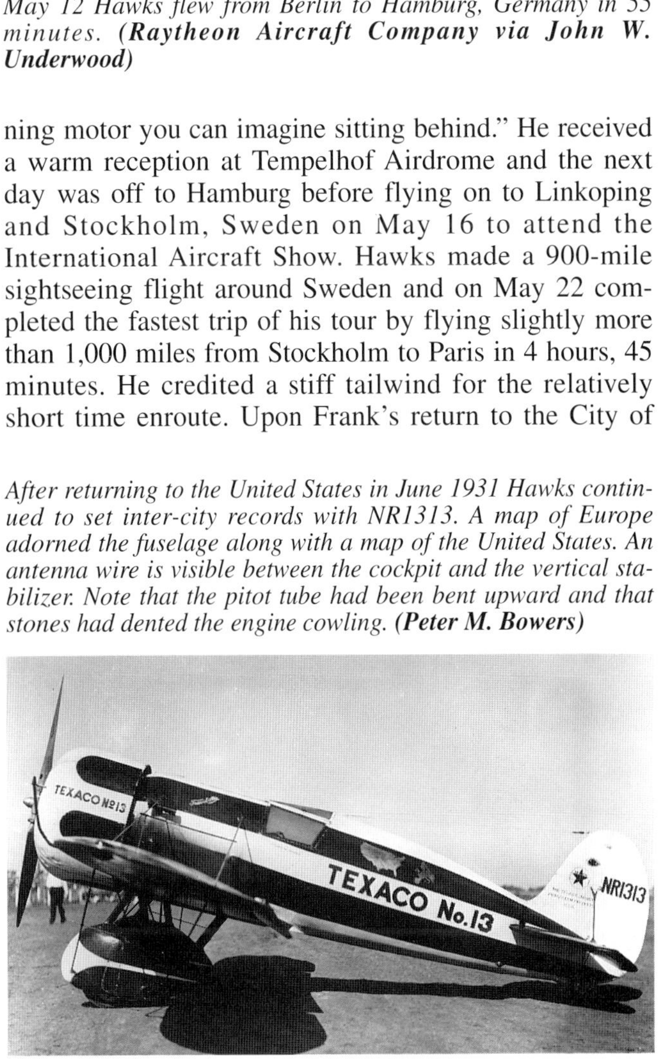

Frank Hawks in the cockpit of Texaco 13. The instrument panel had been reconfigured compared with the original arrangement of the instruments in 1930. ***(Kenneth D. Wilson)***

Lights, Colonel Clifford Harmon, president of the International League of Aviators, declared him the world's outstanding airman for 1930 and presented Hawks with an impressive trophy to commemorate the event, which was held during the French Colonial Exposition.

During the next month Frank kept busy flying throughout Western Europe. One day he flew to London for breakfast, Berlin for lunch, and returned to Paris for a dinner that was followed by an evening at the "Follies." Total flying time for the day was 7 hours, 31 minutes covering a distance of 1,440 miles. With his European jaunt nearing its end, Hawks flew a fast-paced series of trips to Basel, Switzerland, Copenhagen, and Amsterdam. While in Switzerland he demonstrated the ship and met famous World War One German ace Ernst Udet, who was performing aerobatics at an airshow. Throughout his three months on the Continent Hawks spoke with many pilots. He concluded that "they are all good fellows with one common interest at heart—the airplane." One of his most memorable flights was from Lausanne over the Alps on June 9: "I sailed right over the highest peak, Mont Blanc, which is at an altitude of 15,500 feet." On June 10 Hawks was in Rome again having made the 700-mile flight in 3 hours, 31 minutes at an average speed of nearly 200 mph. His wife had left for Rome the previous night and Frank was able to meet her at the train station the next day.

On June 12 he flew from Rome to Torino and from Torino to Milano. The next morning he took off for Budapest, Hungary where he enjoyed a luncheon hosted by the First Secretary to the American Embassy before flying back to Rome. Hawks paid another visit to England on June 15 by flying non-stop from Rome to London in 5 hours, 30 minutes. He then flew back to Rome in 4 hours, 38 minutes and made the return flight to London in only five hours.

As Frank and his wife prepared to board the steamship *"Empress of France"* for their voyage back to the United States, he noted that "A complete box of spare parts for the Wright Whirlwind motor was sent over here and now it is going back with me, unopened. Speaks pretty well for the motor and unquestionably for the Texaco products that were used in that motor during the entire three months over here and 20,000 miles of flying." Although Frank's comments obviously were intended to praise his employer's aviation products, the Wright engine and the Travel Air monoplane had performed well under a wide variety of flying conditions and were deserving of credit.

Hawks, who considered himself a salesman of speed and not a stunt flier bent on intentionally setting records, felt strongly about his self-imposed role. When asked about making speed dashes between cities in America and Europe, Frank replied that he made the flights without running the engine at full throttle, but at a reduced power setting. In June 1931 he described his airplane to

Frank Hawks prepares to board Texaco 13 for a demonstration flight. A colorful Waco Taperwing biplane was parked next to NR1313. ***(Raytheon Aircraft Company)***

The American Magazine: "Some time ago I had a new ship built according to my specifications. The ship can go two hundred fifty, and has a cruising speed of around two hundred. Now, it happens that no other aviator, just at present, has a plane which can cruise at two hundred. Is it my fault, then, if going from one city to another I break a record here and there? Is it my fault, I ask you? I can't help it if my plane goes faster than other planes!"

After Hawks returned to the United States he continued to fly NR1313 and was soon establishing more intercity records. On July 23, 1931 he flew from New York City to Havana, Cuba in eight hours, eight minutes and returned to New York in seven hours 30 minutes, beating a record set earlier that month by James Goodwin Hall. Turning his attention to Canada, on July 31 Frank and his Travel Air dashed between Ottawa and Montreal in a mere 48 minutes, then from Montreal to Toronto in one hour, 48 minutes, Toronto to Ottawa in one hour, 10 minutes, and Ottawa to Montreal in one hour, 10 minutes. He capped off his one-day Canadian marathon by flying from Montreal to New York City in one hour, 45 minutes.

But Hawks was not finished yet. On August 1 he flew from New York to Binghampton in 35 minutes and on August 4 streaked from Wheeling, West Virginia to New York City in one hour, 45 minutes. *Texaco 13* made the trip from New York City to the windy city of Chicago on August 12 in four hours, 4 minutes. Still not satisfied, Frank dashed from Chicago back to New York City on August 14 in 3 hours, 20 minutes followed on August 20 by a long-distance flight from New York City to Fort Worth, Texas that required seven hours, 39 minutes to complete. The next city pair linked by NR1313 were Atlanta and Newark, New Jersey in four hours, 8 minutes on August 26, and a flight on September 23 from Memphis to New York City was completed in only four hours, 50 minutes.

Although these flights had made headlines and brought new fame to Hawks and *Texaco 13,* he already had begun thinking about obtaining a more powerful and faster machine. During the past 12 months the Travel Air Type R had served him well but as with many racing pilots in the early 1930s, Hawks wanted an airplane that could attain or exceed the magic 300 mph. mark. He

A trio of personalities pose with NR1313. Walter Beech (left) Bob Hutchinson of Saskatoon, Saskatchewan, Canada (center), and Frank Hawks. ***(Walter H. and Olive Ann Beech Collection, Ablah Library Special Collections, Wichita State University)***

knew *Texaco 13* was not capable of such performance without extensive modifications, but Hawks also realized that the monoplane still was among the fastest commercial ships in the air.

In November 1931 the speedster underwent a scheduled, semi-annual inspection at the Travel Air factory. New Macwhyte brace wires were installed on the landing gear, and the stabilizer control in the cockpit was reshimmed. Bearings were replaced on the aileron hinges, and the J-6-9 engine received a "top" overhaul that consisted of new pistons, rings, as well as reworking of the cylinder barrels and the valve train components.

With *Texaco 13* refurbished and ready for more aerial adventures, early in December Hawks attempted to break the record between Vancouver, British Columbia and Agua Caliente that had been set by American racing pilot James "Jimmy" Wedell, but failed. Ending the year on a positive note, on December 14 Frank and *Texaco 13* negotiated the 750 miles between Chicago and New York City in a record time of three hours, 15 minutes.

To start the new year off, in January 1932 Hawks flew from Agua Caliente to Oakland, California in two hours,

In April 1932 Hawks had an accident with NR1313 at North Grafton, Massachusetts that ended the airplane's flying career. The Texaco 13 was severely damaged and Hawks was seriously injured. ***(Peter M. Bowers)***

Another view of the airplane after the accident in Massachusetts. Debris had been cleared away in preparation for moving the airplane to a hangar at the North Grafton airport. ***(Smithsonian Institution negative number 83-3563)***

The Texaco 13 was photographed after being rebuilt for static display early in 1933. A note on the photograph reads: "Hello Walter (Beech)—Here's your Mystery S rebuilt as of date. Looks good. Sincerely, Frank." ***(Walter H. and Olive Ann Beech Collection, Ablah Library, Special Collections, Wichita State University)***

23 minutes and then flew from Oakland to Portland, Oregon in two hours, 52 minutes. Next, he dashed between Portland and Vancouver and back again in one hour, 16 minutes and 1 hour, 20 minutes, respectively, and ended his New Year's attack on inter-city records by flying from Portland to Agua Caliente in five hours, 31 minutes. All of these flights were made on the same day, January 23, 1932. Although Hawks did not break the northbound record between Agua Caliente and Vancouver established earlier by Jimmy Wedell, he did beat the southbound record set by James Goodwin Hall by nearly one hour.

In the wake of these record flights and still flushed with the success of his European odyssey in the summer of 1931, early in 1932 Hawks began to plan a series of long-distance record attempts that would take him to key cities and points in the Caribbean Ocean as well as Central and South America. In anticipation of these flights Frank wrote to Gilbert Budwig at the Aeronautics Branch: "I am contemplating a trip to the West Indies and looping on down to Port of Spain and back up to the South American coast, to Panama, Central America, and home." Budwig forwarded the request to the United States' State Department, and on March 22 Clarence M. Young, the Assistant Secretary of Commerce, forwarded the request to the Office of the Secretary of State. The United States Government had no objections to the trip, and officials set about the task of obtaining permission for overflying the sovereign states in Central and South America.

As Hawks began making final preparations for the journey in March 1932, *Texaco 13* was undergoing an inspection for the issuance of another "Restricted" license. At the time of the inspection NR1313 had flown a total of 660 hours since it was delivered to Hawks in July 1930. To provide more range fuel capacity was increased to 130 gallons, and oil capacity to 12 gallons. In addition, the aft fuselage longerons were reinforced and some of the cross-brace tubing was replaced along with the tail skid. The interior of the fuselage was cleaned and flushed, and the engine was changed to another high-compression Wright J-6-9 (serial number 10291). [7]

Hawks and *Texaco 13,* however, would not get the opportunity to make their flight to Port of Spain and beyond. On April 7, 1932 the airplane crashed and was substantially damaged after the engine failed during takeoff from the airport at North Grafton, Massachusetts. Hawks was seriously injured and his face, which had been smashed against the instrument panel, required reconstructive plastic surgery. He spent weeks recovering in the hospital. Meanwhile, the wrecked monoplane was dismantled and stored in a hangar at the airport. Hawks was undecided about whether to rebuild the ship or dispose of it.

On July 19, Blanche N. Gallagher, Frank's personal secretary at The Texas Company, wrote to Robert R. Reining at the Registration Section of the Aeronautics Branch: "The *Texaco 13* is now at Ames-Skyways in Boston, Massachusetts where it is being repaired for

exhibition purposes. Since it is the intention to use it hereafter merely as an exhibit, I do not believe that a renewal of license NR-1313 is desired."

As a result, the license was suspended in July. In November 1932 as reconstruction work was beginning, The Texas Company requested that NR1313 be reassigned to the airplane. It was scheduled to be placed on public display when the reconstruction was completed. The Aeronautics Branch granted the request, and *Texaco 13* was eventually donated to the Museum of Science and Industry (also known as the Rosenwald Museum) in Chicago, Illinois and placed on permanent display.

"TEXACO 13" SPECIFICATIONS
(Typical configuration)

Wing span (standard)	30 feet 6 inches
Wing span (short, one panel)	11.6 feet
Wing chord (standard)	63 inches
Wing chord (short)	60 inches
Wing area (standard)	140 sq. ft.
Wing area (short)	125 sq. ft.
Wing loading (standard)	18.2 lbs./sq. ft.
Wing loading (short)	22 lbs./sq. ft.
Length	20 feet 2 inches
Height	7 feet 9 inches
Empty weight	1,880 pounds
Gross weight (standard wings)	Approx. 2,800 pounds
Gross weight (short wings)	2,742 pounds
Engine	Wright J-6-9 400-465 hp.
Fuel capacity	109 gallons
Useful load	400-465 pounds
Maximum speed	250 mph.
Maximum range	1,000 statute miles
Diving speed	300 mph.
Rate of climb	3,200 ft./minute

The Texaco 13 eventually was placed on display at the Museum of Science and Industry in Chicago, Illinois. ***(Kenneth D. Wilson)***

Although *Texaco 13* would fly again only in the imaginations of those who appreciated her graceful form, there was consolation in the knowledge that Frank Hawks and The Texas Company had saved a famous airplane from destruction and preserved it for future generations to admire.

In a partnership between man and machine that lasted fewer than 21 months, the speed team of Frank Monroe Hawks and *Texaco 13* had earned a special place not only in the history of aviation, but in the annals of the Travel Air Type R monoplanes and their many accomplishments.

CHAPTER SIX NOTES

1. Letter from Walter H. Beech to Gilbert Budwig dated July 9, 1930.
2. Walter L. Pike was a regional inspector with the Aeronautics Branch based at Kansas City, Missouri.
3. Hawks used the "Moto-vita" to help him lean the mixture and conserve fuel during the transcontinental flights.
4. By 1930 the popular Wright J-6-9 was producing 450 hp. and special modifications further increased power. It proved to be a highly reliable powerplant.
5. When Hawks landed in New York 25 gallons of remained fuel in the tanks.
6. The short-span wings for Texaco 13 eventually were donated to the Staggerwing Museum Foundation in Tullahoma, Tennessee and placed on permanent display.
7. The engine change was recorded in Department of Commerce documents dated March 25, 1932.
8. Gardner, Lester D.: *"Who's Who In American Aeronautics."* Aviation Publishing Corporation, New York, N.Y., 1928.
9. *"Popular Aviation,"* October 1938. Hawks held Transport Pilot's license No. 498 and Federation Aeronautique Internationale license No. 4493.

The Italian Type R

R-2005 - 11717 AND MM185

Early in the summer of 1931 the Curtiss-Wright Export Corporation received an order from the Italian government for a Type R monoplane. Specifically, the order was from the Regia Aeronautica (Italian Royal Air Force) and a contract for the airplane was signed on June 10 by Lieutenant Colonel Paolo Sbernadori, Italian Air Attache to the United States. The cost of the airplane was not to exceed $21,000. [1]

Italy's interest in the Type R had been ignited in the summer of 1931 when Frank Hawks toured the European Continent for three months. He made a number of flights to Rome where his *Texaco 13* was demonstrated and made available for examination by civilian and military pilots. Among those who appreciated the ship's sleek lines as well as its performance was General Italo Balbo, Air Minister to Italian dictator Benito Mussolini. In January 1931 Balbo had gained international attention by leading a flight of 12 Savoia-Marchetti S.55 seaplanes on a transatlantic flight to Brazil. During July and August 1933 he led a flight of 24 Savoia-Marchetti S.55X aircraft on a 12,000-mile, 45-day aerial escapade from Italy to Chicago, Illinois and return. An experienced airman and fighter pilot, Balbo recognized the Travel Air Type R's potential for influencing the design of future fighter aircraft in Italy. [2]

But the order posed an immediate problem for Walter Beech. The company had officially terminated the Type R after it completed the *Texaco 13* for Frank Hawks in July 1930. The Italian sale, however, represented new business that was becoming increasingly difficult to find as the Great Depression tightened its merciless grip on America in 1931. The once-thriving Travel Air factory in Wichita had ceased operations as a manufacturing facility and

Newman Wadlow was one of two pilots assigned by Walter Beech to perform test flights on R-2005 in July 1931. The airplane was nearly identical to the Texaco 13. ***(Carl Burnham)***

Detailed photograph shows four-piece cowling with internal bracing as well as landing gear and paint design. The small hole in the right side of the cowling accepted a hand crank for the inertia starter. ***(Carl Burnham)***

was being used chiefly to store airplane parts and assemblies. To further compound Beech's problem, the majority of Travel Air's workforce had been furloughed as production of airplanes continued to decline. He would have to find skilled workers to build the Type R, and more importantly, direct Herbert Rawdon and Walter Burnham to supervise the entire project and deliver the ship on time and at the specified price.

Walter Beech contacted Rawdon and discussed the Italian order. In his role as a Curtiss-Wright senior executive Beech was too busy flying a desk to become directly involved in the project, but it ultimately was his responsibility to ensure that the job was done. Herb Rawdon and Walter Burnham would be responsible for the airplane's construction and its flight test program, and Beech emphasized to both men that there was no time to waste as the Italians expected delivery in only two months.

Rawdon and Burnham began the process of ordering components and parts, instruments, materials, the high-compression Wright J-6-9 radial engine and a Hamilton Standard steel propeller. Burnham was placed in overall charge of construction. A team of 20 men was recruited to build R-2005 and the majority were the same craftsmen who had built the previous four Type R monoplanes. Theodore "Ted" Cochran led the woodworking crew and Howard Bacchus was overall foreman and supervised welding operations. Work on the wings began June 12 and was completed on June 24, except for final painting. Fuselage tubing and heat treatment of specific tubes, including the engine mount, had been completed by June 23 when the engine (serial number 12507) was scheduled to arrive. It was installed on or about June 27, and by June 29 the fuselage structure was complete and partially painted, as were the control surfaces. The landing gear shock absorbers and slides were welded and machined by June 18, and the fuel tanks, NACA cowling, and other streamline fairings were completed by July 10, 1931. [3]

In addition, the following cockpit instrumentation and equipment was ordered:

1. Pioneer compass, rate of climb indicator and turn and bank indicators, 250-mph. airspeed indicator, altimeter, fuel and oil pressure gauges, 2,500 RPM. tachometer.
2. Jaeger stop watch.
3. J.W.C. Schenck & Company artificial horizon and bank indicator.
4. Type 127 instrument board lights.
5. Fifty square feet of Red Aero Leather for the pilot seat and cockpit interior.

The airplane was painted in a red and white scheme similar to that on *Texaco 13* except that the vertical stabilizer featured three broad stripes painted in the national colors of Italy. In addition, the numbers "11717" were

The Regia Aeronautica's Type R photographed with one segment of the cowling removed. The view illustrates how the redesigned cowling improved access to the engine compartment compared with the two-piece design used on R614K, and originally installed on NR482N. ***(Carl Burnham)***

Comprehensive view of the cockpit of the Italian airplane shows the control stick, instrument panel, throttle, spark, and mixture controls, fuel selector, magneto switch, fuel sight gauge, as well as Pyrene fire extinguisher (lower left) designed to discharge into the engine compartment. The main fuel tank is visible forward of the control stick, as are fuse blocks above the extinguisher, and the Burgess dry cell battery that powered instrument panel lights. Six electrical switches are installed on left side of the panel. A hand-held fire extinguisher was mounted on a steel tube. **(Raytheon Aircraft Company)**

temporarily assigned by the Aeronautics Branch and applied to the vertical stabilizer as a temporary identification to allow Travel Air to flight-test the aircraft for export purposes. According to Herb Rawdon's engineering notes, R-2005 had a fuel capacity of 108 gallons and an oil capacity of 8 gallons.

By mid-July the Italian Type R was nearly ready to make its first flight. Although Walter Beech would have preferred that Clarence Clark fly the ship, Clark had left Travel Air in 1930 to fly for Phillips Petroleum in Bartlesville, Oklahoma. Beech had contacted Phillips Petroleum's chief pilot, Billy Parker, and asked that Clarence be granted leave for about one week to test-fly the airplane. Parker refused. With time slipping away, Beech talked with Rawdon and tapped young Newman Wadlow and Wichita pilot L.G. "Swede" Larson to share pilot duties on R-2005. Newman and his identical twin brother Truman Wadlow operated a flying school on the east side of Travel Air Field. Both young men had been taught to fly when they began working for Walter Beech in 1925 and he had arranged for them to receive flying lessons in return for their labor.

Larson was first to fly the new ship. According to Rawdon's notes, on July 19, 1931 Larson made two flights from Travel Air Field that totaled 45 minutes. The engine was operated at 2,300 RPM. and no major problems were encountered. On July 21 Larson made three flights and logged one hour, and on July 22 he flew the ship again but only logged 20 minutes.

Newman Wadlow took over from Larson later that day and flew R-2005 for 2 hours, 35 minutes, followed on July 23 by a unprecedented 17 flights totaling 4 hours, 50 minutes. The tests were necessary to help Walter Burnham identify and eliminate a vibration in the

Cockpit of R-2005 after completion of the airplane. Left side panel for the canopy enclosure is open. **(Theodore Cochran)** ▼

L.G. Larson was photographed flying the monoplane east of Travel Air Field on its first flight, July 19, 1931. Department of Commerce identification number 11717 has been applied to the vertical stabilizer but was removed when the airplane was shipped to Italy. R-2005 was equipped with standard wings. **(Kenneth D. Wilson)**

empennage that had been encountered earlier that day when the airplane was flown at full throttle. In 1981 Wadlow told the author during an interview that Burnham had the cowling removed and the ship flown, but the problem persisted. After several more flights had been made without identifying the cause, Newman proposed that oil be poured on the aft fuselage and empennage to reveal turbulence. His suggestion was accepted and he soon took off again. After Newman landed, Burnham scrutinized the oil patterns and concluded that the culprit was a deformity in the wing trailing edge. Apparently, during construction the shape of the trailing edge did not conform precisely to requirements and induced turbulence into the airflow coming off the wing. The air struck the empennage surfaces and at high speeds and induced a vibration into the structure. The trailing edge was reworked and the problem disappeared.

Five additional flights were made to test the airplane. These included a final, 25-minute test on July 24, and four flights on July 25 totaling one hour, 35 minutes. After 9.7 hours of flying, Rawdon and Burnham pronounced R-2005 ready for delivery. The airplane was rolled into the factory and thoroughly cleaned in preparation for a visit and official acceptance of the ship by Italian Air Attache Lieutenant Colonel Paolo Sbernadori, scheduled for late in July. Sbernadori traveled to St. Louis and flew to Wichita with Walter Beech. Although Sbernadori was sick with ptomaine poisoning he insisted on make the trip and was anxious to both examine and fly R-2005 as a representative of his government.

The illness, however, precluded any flights during his brief stay in Wichita. After determining that the airplane met specifications stipulated in the purchase order,

The Italian Travel Air Type R influenced Breda's design of the Ba. 27 fighter. A small number were sold to China in the mid-1930s. **(Gianni Gambarini)**

Front, quarter view of MM185 when it was assigned to the fighter pilot training school at Aviano, Italy. The airplane was dismantled and scrapped in 1937. The Italian words at the bottom of the photograph, "Apparecchio de caccia americano Texano," translates as "Texan American fighter aircraft—a reference to the Texaco 13 upon which MM185 was based. ***(Roberto Gentilli)***

Sbernadori praised the airplane and its virtues to local reporters. He departed Wichita and a few days later the machine was dismantled and carefully packed into two large crates for shipment to Italy. When it arrived in August technicians assembled the aircraft using specific instructions provided by the Travel Air factory.

Designated as *"MM 185"* by the Regia Aeronautica as an experimental aircraft, the ship was not assigned a civilian registration number. According to Italian aviation historian Gianni Gambarini, beginning in August 1931 the ship was flown extensively by the Reparto Sperimentale di Volo (experimental flight unit) based at Montecelio near Rome. In 1932 *MM 185* was assigned to the Scuola Caccia (fighter school) at Aviano where it apparently remained for the next five years. In May 1937, however, the airplane was reassigned to the Reparto Sperimentale di Volo and was dismantled and scrapped.

The Travel Air Type R did inspire one Italian fighter design, the Breda Ba. 27 low-wing monoplane. Two prototypes were built initially (MM 217 and MM 218), followed by a third aircraft that was powered by a Bristol *Mercury VI* radial engine built under British license by Alfa-Romeo. Although no orders were received from the Regia Aeronautica, a small batch of production airplanes were purchased by the Chinese government in the mid-1930s.[4]

CHAPTER SEVEN NOTES

1. Lieutenant Colonel Sbernadori had been a reconnaissance pilot in World War One.
2. *New York American* newspaper, July 20, 1933 and *Aero Digest* magazine, August 1933.
3. According to Herbert Rawdon's engineering notes, each wing (with aileron) cost $2,040, regardless of whether it had 125 square feet of area or 140 square feet of area.
4. Letter to the author from Gianni Gambarini dated January 10, 1999.

The Italian Type R was photographed soon after it arrived in Italy in August 1931. It was assigned military designation MM185 and was flown extensively by the Regia Aeronautica. ***(Roberto Gentilli)***

TYPE R SPECIFICATIONS

The information presented is based on Rawdon's compilation for the Type R racing monoplane dated October 10, 1929. The figures represent "Final or observed" results obtained with the airplane as it was configured for the National Air Races, and differ from original or preliminary estimates.

General Design Data

Designation	*Computed*	*Final*
Wing Span	30 feet 6 inches	29 feet 2 inches
Length (overall)		20 feet 2 inches
Height (overall)		7 feet 9 inches
Wing chord	60 inches	60 inches
Total wing area (sq. ft.)	141	125
Aileron area (sq. ft.)		12.3
Stabilizer area		14.1
Elevator area		9.4
Rudder area		5.7
Vertical stabilizer area		4.1
Wing loading (lbs./sq. ft.)	12.05	15.5
Power loading (lbs./HP)	6.8	4.6
Gross weight	1,700 lbs.	1,940 lbs.
Empty weight	1,278 lbs.	1,475 lbs.
Useful load	422 lbs.	465 lbs.
Weight (stress analysis)	1,750 lbs.	1,750 lbs.
Engine horsepower	250 hp.	425 hp.
Propeller diameter		98 inches
Wheel tread		68 inches
Wing aspect ratio	6.25	6.26
Tail surface aspect ratio		2.84
Aileron aspect ratio		5.85
Wing section	M-6	RAF 34
Angle of incidence	2.25	1 3/4
Dihedral angle		4.0
Landing angle		13 degrees

Performance

Maximum speed	165 mph.	227 mph.
Stall speed	68 mph.	70 mph.
Landing speed	65 mph.	70 mph.
Rate of climb (ft./min.)	1,850	3,000
Climb to 1,000 ft.		34 sec.
Climb to 5,000 ft.		1.85 min.
Climb to 10,000 ft.	7.53 min.	4.05 min.
Service ceiling	20,800 ft.	30,000 ft.
Time to service ceiling	34.5 min.	35.8 min.
Absolute ceiling	22,000 ft.	31,000 ft.
Gliding angle		1:9.1
Range	501 mi.	525 mi.
Cruising speed		150 mph.

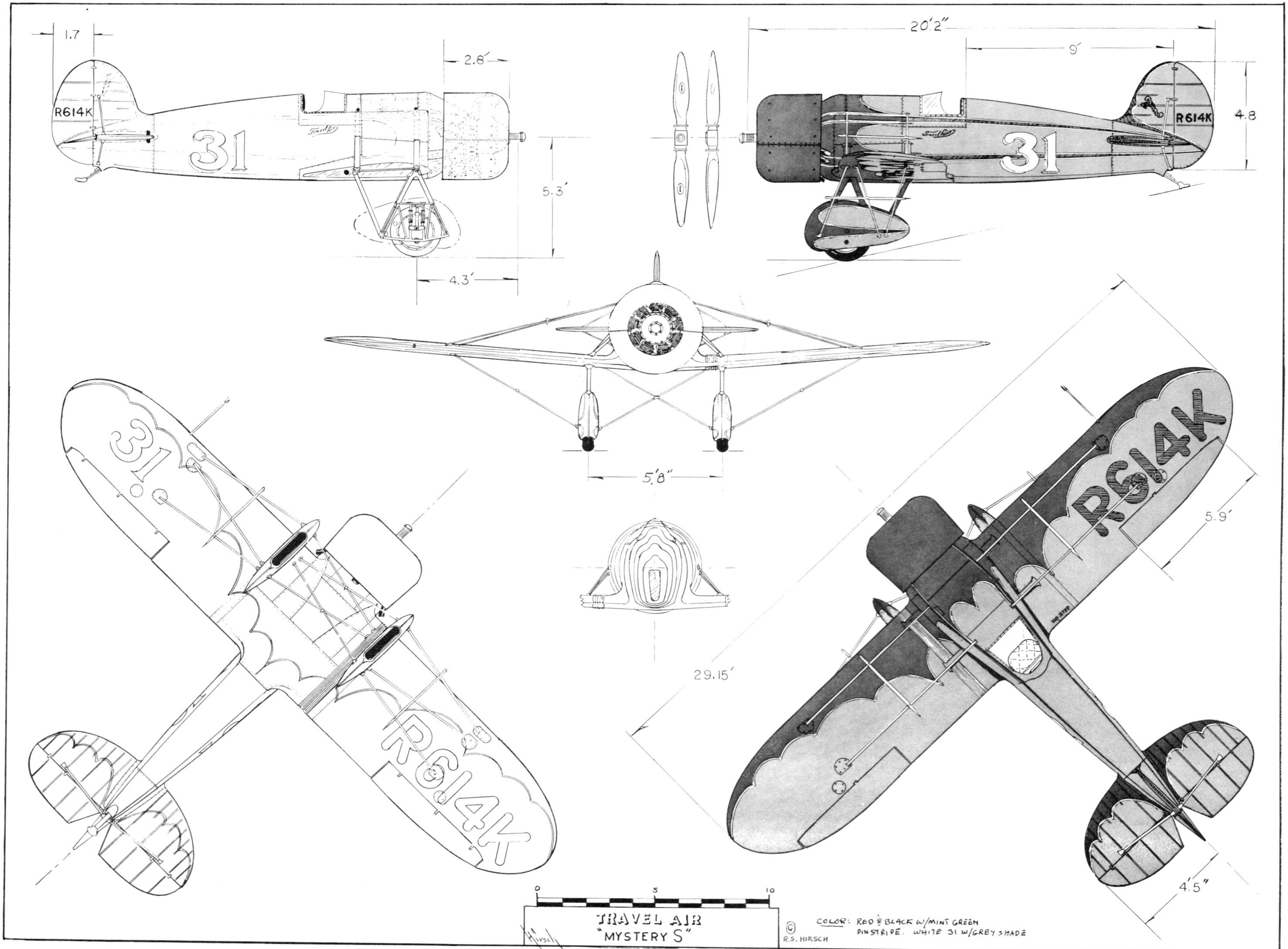
1.7
2.8'
R614K
31
5.3'
4.3'
20'2"
9'
4.8
5'8"
5.9'
29.15'
4'5"
0
5
10
TRAVEL AIR
"MYSTERY S"
© R.S. HIRSCH
COLOR: RED & BLACK W/MINT GREEN PINSTRIPE. WHITE 31 W/GREY SHADE

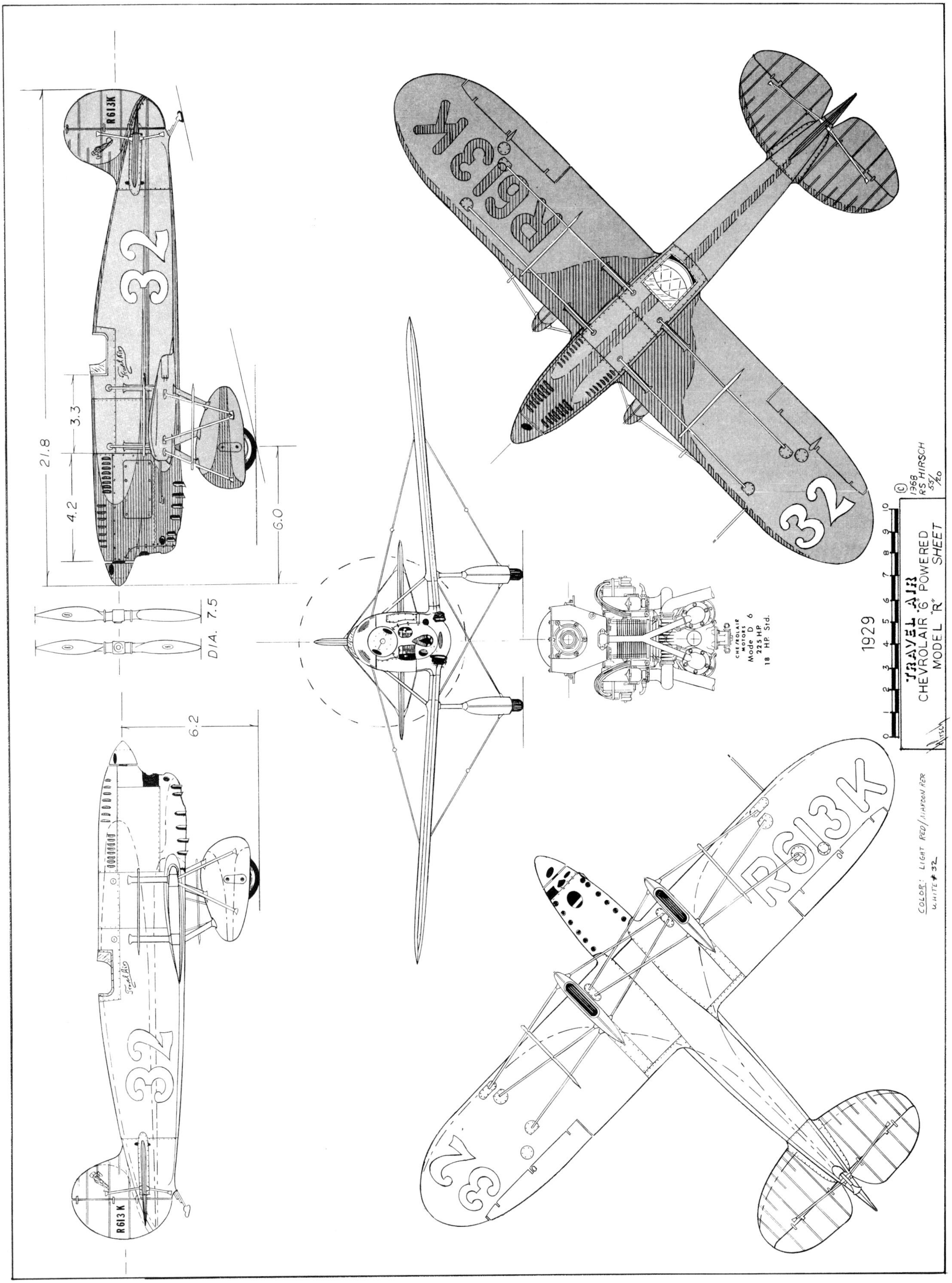

TRAVEL AIR
CHEVROLAIR "6" POWERED
MODEL "R" SHEET
1929
© 1968 RS HIRSCH
COLOR: LIGHT RED/MAROON RED
WHITE #32
CHEVROLAIR MOTORS
225 HP
R613K
32
21.8
3.3
4.2
6.0
6.2
DIA. 7.5

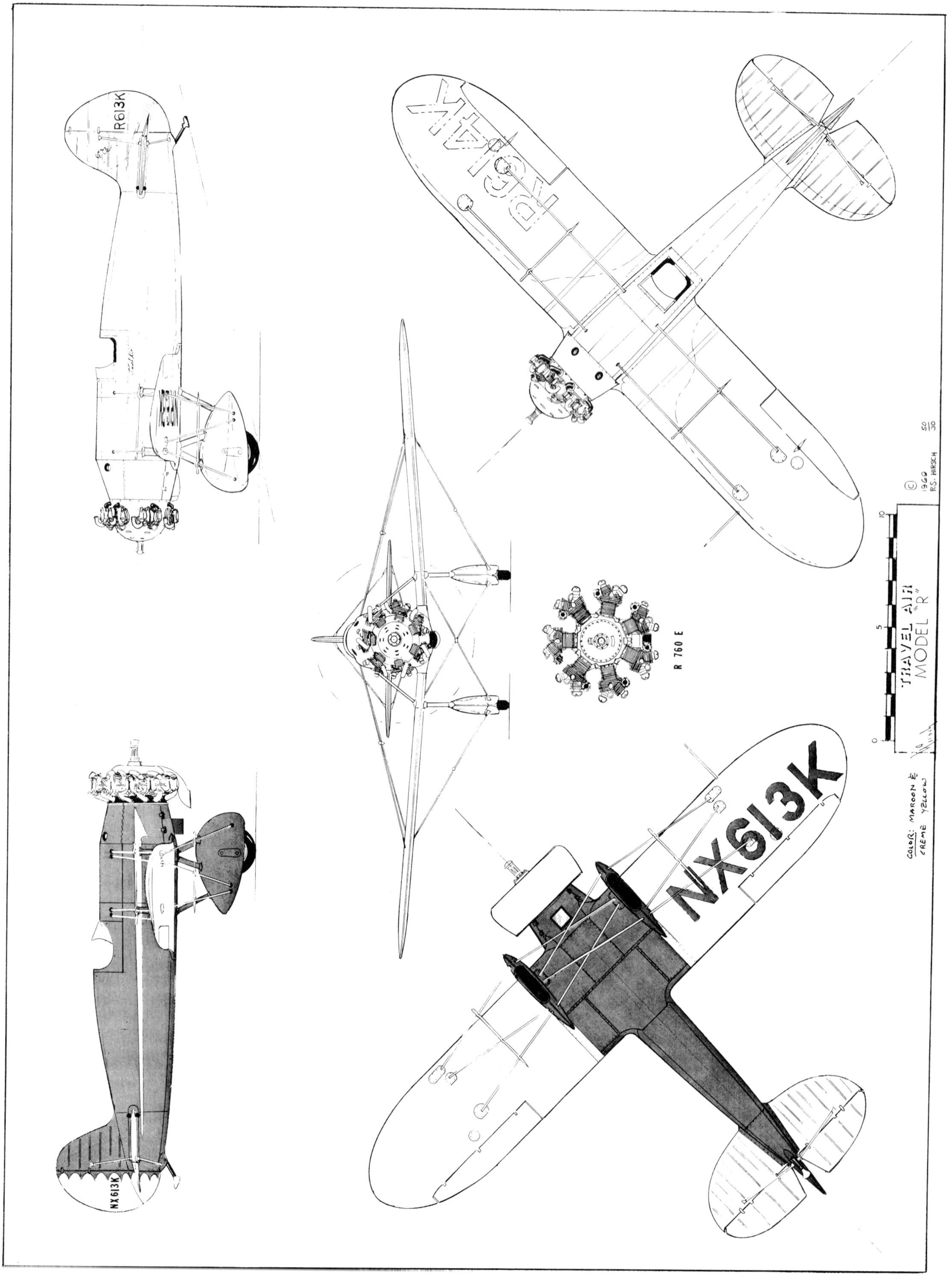
R613K
NX613K
R 760 E
TRAVEL AIR
MODEL "R"
0
5
10
© 1969 R.S. HIRSCH
50/30
COLOR: MAROON & CREME YELLOW

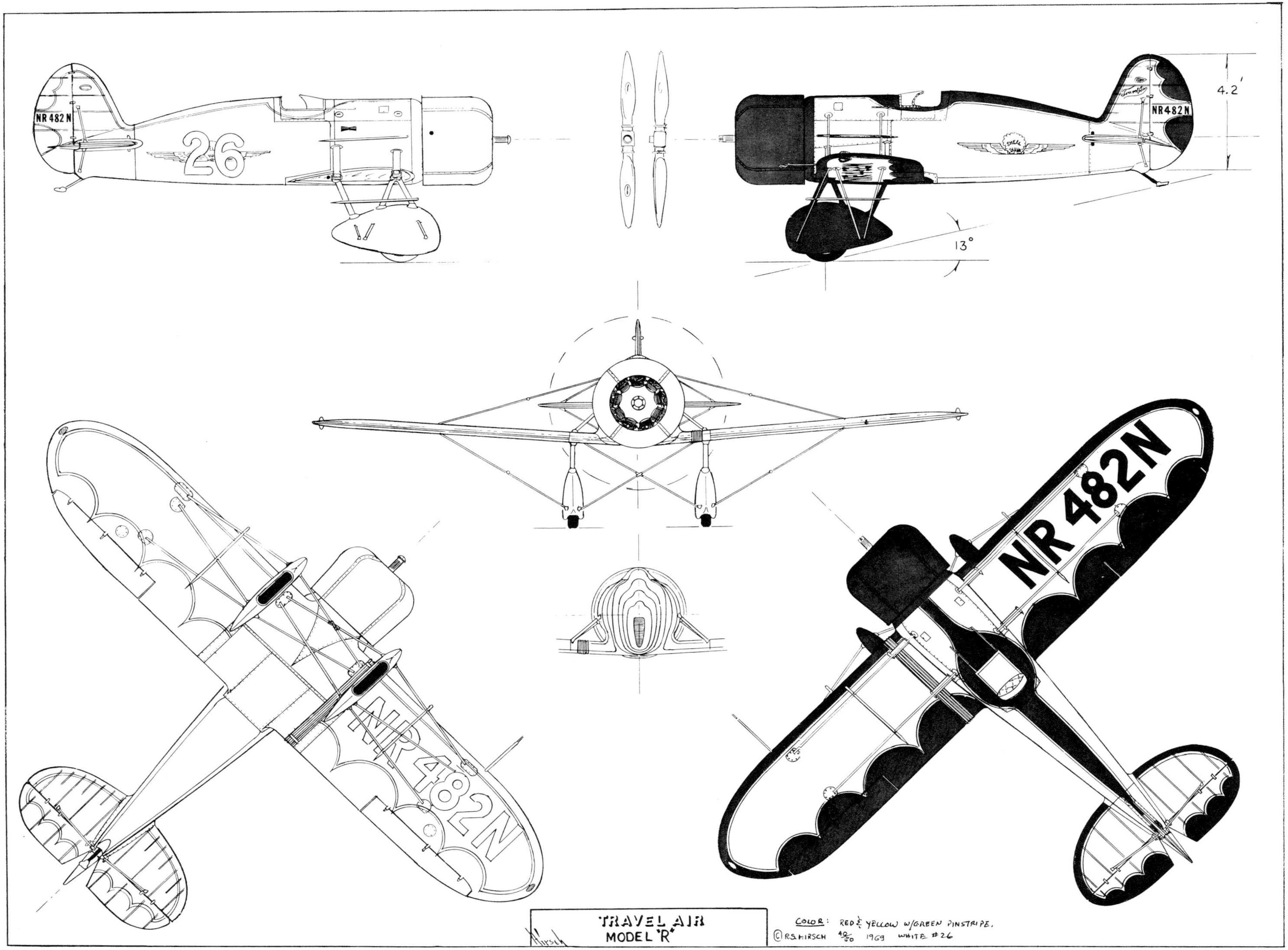
NR482N
26
4.2'
13°
NR482N
TRAVEL AIR
MODEL 'R'
COLOR: RED & YELLOW W/GREEN PINSTRIPE.
WHITE #26
©R.S.HIRSCH 1969
Hirsch

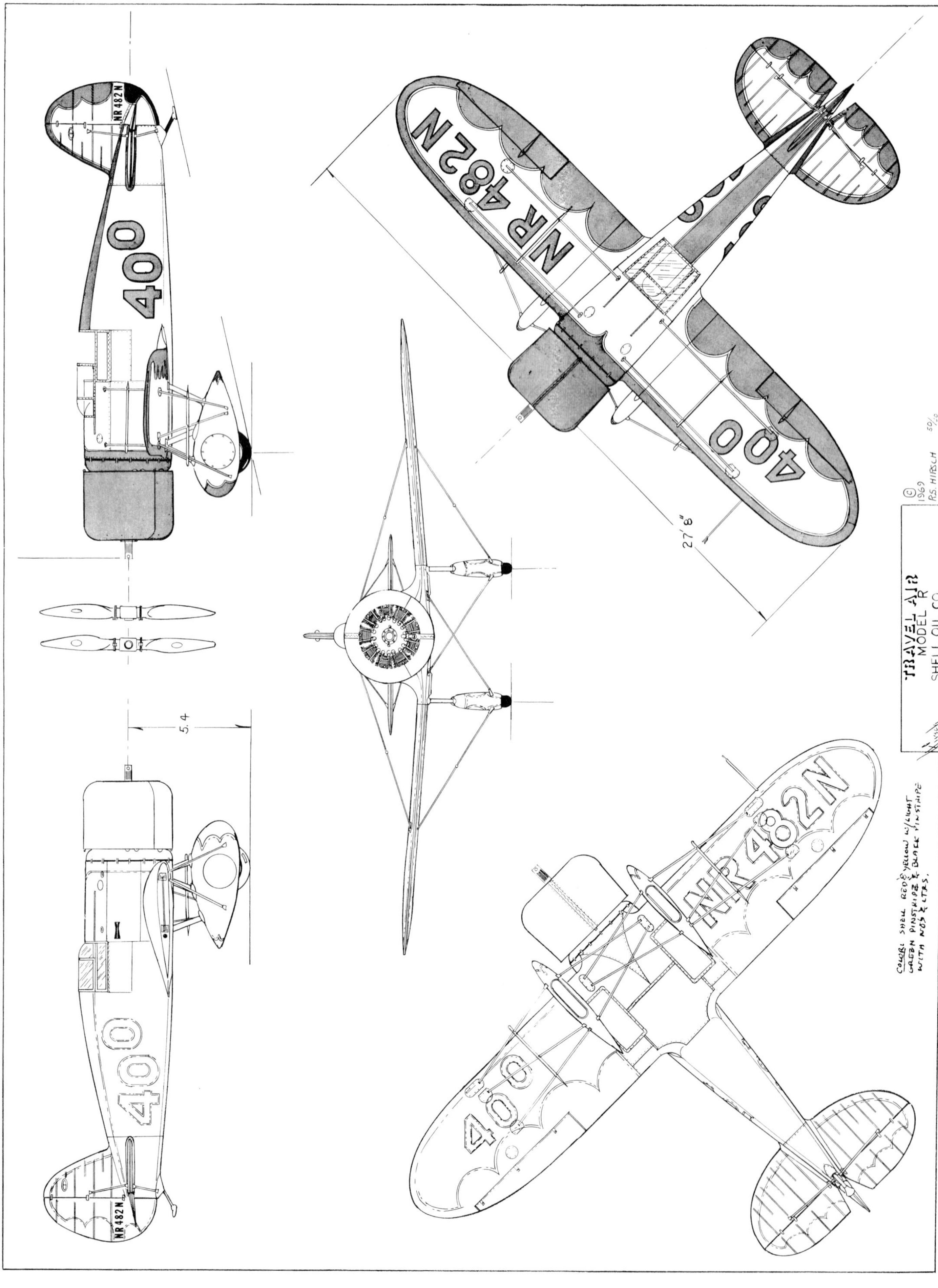
NR482N
400
27'8"
5.4
TRAVEL AIR
MODEL "R"
SHELL OIL CO.
© 1969 R.S. HIRSCH
COLOR: SHELL RED & YELLOW W/ LIGHT
GREEN PINSTRIPE & BLACK PINSTRIPE
WITH NOS & LTRS.

1st COLOR: RED (DEEP) & TAN (LIGHT)
TEXACO No 13
NR1313
TEXACO 13
NR1313
DIA 7.9
4.2
5.1
5°
30.3
1930-31
TRAVEL AIR
MODEL 'R'
FRANK HAWKS
SHEET 4
© 1965

BIBLIOGRAPHY

In addition to original research by the author, the following publications served as references in the preparation of this book:

1. Allen, Richard Sanders: *Revolution In The Sky.* Orion Books, New York, N.Y., 1988.
2. Berliner, Don: *Worldwide Directory of Racing Airplanes.* Aviation Publishing, Inc., Destin, Florida, 1997.
3. Bowers, Peter M.: *Curtiss Aircraft - 1907-1947.* Putnam & Company, Ltd., London, England, 1979.
4. Christy, Joe and Shamburger, Page: *The Curtiss Hawks.* Wolverine Press, Kalamazoo, Michigan, 1972.
5. Forden, Leslie: *The Ford Air Tours - 1925-1931.* The Nottingham Press, Alameda, California, 1973.
6. Glines, Carroll V.: *Jimmy Doolittle - Daredevil Aviator And Scientist.* The Macmillan Company, New York, N.Y., 1972.
7. Hawks, Frank M.: *Speed!* Brewer, Warren and Putnam, New York, N.Y., 1931.
8. Munson, Kenneth: *Airliners Between The Wars - 1919-1939.* The Macmillan Company, New York, N.Y., 1972.
9. Phillips, Edward H.: *Travel Air - Wings Over The Prairie.* Flying Books International, Eagan, Minnesota, 1982.
10. Phillips, Edward H.: *The Staggerwing Story.* Flying Books International, Eagan, Minnesota, 1996.
11. Schmid, S.H. and Weaver, Truman C.: *The Golden Age of Air Racing.* Experimental Aircraft Association, 1991.
12. Schultz, Barbara Hunter: *Pancho - The Biography of Florence Lowe Barnes.* Little Buttes Publishing Company, Lancaster, California, 1996.
13. Smith, Herschel: *Aircraft Piston Engines.* McGraw-Hill, Inc., New York, N.Y., 1981.
14. Thomas, Lowell and Jablonski, Edward: *Doolittle - A Biography.* Doubleday & Company, Inc. Garden City, N.Y., 1976.
15. Newspaper accounts, *The Wichita Eagle.* June 1929-December 1931.

ABOUT THE AUTHOR

Edward H. Phillips is a long-time admirer and aficianado of the Travel Air Company, the airplanes it produced, and the men and women who designed, built, and flew them. A resident of Texas, Phillips writes for a major international aerospace publication and also contributes free-lance articles on aviation history to a number of magazines.

In addition to being a historian specializing in general aviation subjects, he is an active commercial pilot, flight instructor, and airframe and powerplant technician. Phillips has written six other books that chronicle the history of key companies and benchmark airplanes, including *Travel Air, the Beech Aircraft Corporation* (now Raytheon Aircraft Company); the *Beechcraft Model 17 "Staggerwing," Piper Aircraft Corporation*, and the *Cessna Aircraft Company.*

INDEX